APPLIED DRILLING CIRCULATION SYSTEMS

APPLIED DRILLING CIRCULATION SYSTEMS

HYDRAULICS, CALCULATIONS, AND MODELS

BOYUN GUO, Ph.D.
University of Louisiana at Lafayette

GEFEI LIU
Pegasus Vertex, Inc.

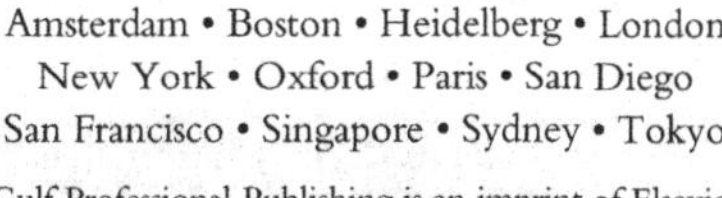

Gulf Professional Publishing is an imprint of Elsevier
30 Corporate Drive, Suite 400, Burlington, MA 01803, USA
The Boulevard, Langford Lane, Kidlington, Oxford, OX5 1GB, UK

Notices

Knowledge and best practice in this field are constantly changing. As new research and experience broaden our understanding, changes in research methods, professional practices, or medical treatment may become necessary.

Practitioners and researchers must always rely on their own experience and knowledge in evaluating and using any information, methods, compounds, or experiments described herein. In using such information or methods they should be mindful of their own safety and the safety of others, including parties for whom they have a professional responsibility.

To the fullest extent of the law, neither the Publisher nor the authors, contributors, or editors, assume any liability for any injury and/or damage to persons or property as a matter of products liability, negligence or otherwise, or from any use or operation of any methods, products, instructions, or ideas contained in the material herein.

Library of Congress Cataloging-in-Publication Data

Guo, Boyun.

Applied drilling circulation systems : hydraulics, calculations, and models / Boyun Guo, Gefei Liu.

p. cm.

Includes bibliographical references and index.

ISBN 978-0-323-28194-2 (paperdback)

1. Mud pumps. 2. Oil-well drilling. 3. Gas well drilling. I. Liu, Gefei. II. Title.

TN871.27.G86 2011

622'.3381–dc22 2010043249

British Library Cataloguing-in-Publication Data

A catalogue record for this book is available from the British Library.

For information on all Gulf Professional Publishing publications visit our Web site at *www.elsevierdirect.com*

Printed in the United States

11 12 13 14 15 10 9 8 7 6 5 4 3 2 1

To

Huimei Wang

An Exemplary Wife and Mother

CONTENTS

PREFACE

Drilling circulation systems in the oil and gas industry have advanced significantly in the last decade. The major changes resulted from the merging of air and gas drilling and underbalanced drilling with traditional liquid drilling systems. During the several years of teaching drilling engineering courses in both academia and industry, the authors realized the need for a book that covers modern drilling practices. The books that are currently available fail to provide adequate information about how engineering principles are applied to solving problems that are frequently encountered in drilling circulation systems. This fact motivated the authors to write this book.

This book is written primarily for well drilling engineers and college students of both senior and graduate levels. It is not the authors' intention to simply duplicate general information that can be found in other books. This book gathers the authors' experiences gained through years of teaching courses on drilling hydraulics, air and gas drilling, and underbalanced drilling engineering in the oil and gas industry and in universities. The mission of this book is to provide drilling engineers with handy guidelines for designing, analyzing, and operating drilling circulation systems.

This book covers the full scope of drilling circulation systems. Following the sequence of technology development, it contains ten chapters presented in three parts.

Part I contains four chapters that cover liquid drilling fundamentals as the first course for entry-level drilling engineers and undergraduate students. Chapter 1 presents an introduction to the equipment used in mud drilling systems. Chapter 2 documents mud hydraulics fundamentals that are essential for selecting and optimizing mud pumps and mud hydraulics programs. Chapter 3 covers in detail the procedure for selecting mud pumps. Chapter 4 presents techniques used to optimize mud hydraulics programs.

Part II includes three chapters that present the principles and rules of designing and operating gas drilling systems. Chapter 5 introduces the equipment used in gas drilling systems. Chapter 6 covers in detail the procedure for selecting gas compressors. Chapter 7 provides guidelines for operating gas drilling systems.

Part III consists of three chapters that cover underbalanced drilling systems. Chapter 8 presents an introduction to the equipment used in underbalanced drilling systems. Chapter 9 describes the procedure for selecting gas and liquid flow rates. Chapter 10 provides guidelines for underbalanced drilling operations.

Since the substance of this book is virtually boundless in depth, knowing what to omit was the greatest challenge. The authors believe that it requires many books to describe the foundation of knowledge in drilling circulation systems. To counter any deficiency that might arise from space limitations, this book provides a reference list of books and papers at the end of each chapter so readers can consult other sources of information about the topics discussed.

As for presentation, this book focuses on providing and illustrating engineering principles used for designing and optimizing drilling circulation systems, rather than on discussion of in-depth theories. Derivation of mathematical models is beyond the scope of this book. Applications of the principles are illustrated by solving example problems. While the solutions to some simple problems that do not involve iterative procedures are demonstrated with stepwise calculations, complicated problems are solved with computer spreadsheet programs, some of which can be downloaded from the publisher's website. The combination of the book and the computer programs creates the perfect tool kit for drilling engineers for performance of their daily work in the most efficient manner. All of the computer programs were written in the form of spreadsheets in MS Excel that are available in most computer platforms in the oil and gas industry. These spreadsheets are accurate and very easy to use. While both U.S. field units and SI units are used in the book, the option of using U.S. field units or SI units is provided in the spreadsheet programs.

This book is based on numerous documents, including reports and papers accumulated through years of work at the University of Louisiana at Lafayette and at Pegasus Vertex, Inc. The authors are grateful to the university and the company for permission to use these materials. Special thanks go to Chevron USA for providing professorships in petroleum engineering throughout the editing of this book. Our thanks also go to Mr. Guoqiang Yin of Shell USA, who contributed a thorough review and edit of this book. On the basis of the collective experiences of the authors and of reviewers, we expect this book to be of great value to drilling engineers in the oil and gas industry.

Dr. Boyun Guo
Chevron Endowed Professor in Petroleum Engineering
University of Louisiana at Lafayette

PART ONE

Liquid Drilling Systems

Liquid is widely used as a circulating fluid to drill oil and gas wells, water wells, geotechnical boreholes, and mining boreholes. This type of drilling fluid is most commonly suspensions of clay and other materials in water and is thus called *drilling mud*. Since it usually has a density ranging from 8.33 to 18.33 ppg (1.0–2.2 S.G.), drilling mud is used for drilling rock formations with normal to abnormal fluid pressure gradients (>0.433 psi/ft or 0.01 MPa/m). Because the bottomhole pressure is greater than the formation pore pressure, mud drilling is an overbalanced drilling operation. Compared to gas drilling and two-phase drilling (see Parts II and III), mud can be used to drill various types of formation rocks with better control of formation fluids and borehole stability.

Part I provides drilling engineers with basic knowledge about mud circulating systems and techniques that are used for optimizing drilling hydraulics to achieve the maximum drilling rate. Materials are presented in the first four chapters:

Chapter 1: Equipment in Mud Circulating Systems
Chapter 2: Mud Hydraulics Fundamentals
Chapter 3: Mud Pumps
Chapter 4: Mud Hydraulics Optimization

CHAPTER ONE

Equipment in Mud Circulating Systems

Contents

1.1 INTRODUCTION

Figure 1.1 shows a typical mud circulating system (Lyons et al., 2009). The drilling mud travels (1) from the steel tanks to the mud pump; (2) from the pump through the standpipe and the kelly to the drill string, which consists of drill pipes and the bottomhole assembly (BHA), with a major length of drill collars; (3) through the drill string to the bit; (4) through the nozzles of the bit and up the annular space between the drill string and the borehole (open hole and cased hole sections) to the surface; and (5) through the contaminant-removal equipment back to the suction tank. The contaminant-removal equipment can include shale shakers, degassers, hydrocyclones (desanders and desilters), and centrifuges. An integrated unit of desanders and desilters is called a mud cleaner. This chapter provides a brief introduction to the equipment that controls the circulating pressure of the system.

1.2 MUD PUMPS

Mud pumps serve as the heart of the mud circulating system. Reciprocating piston pumps (also called slush pumps or power pumps) are widely used for drilling oil and gas wells. The advantages of the reciprocating positive-displacement pump include the ability to move high-solids-content fluids laden with abrasives, the ability to pump large particles, ease

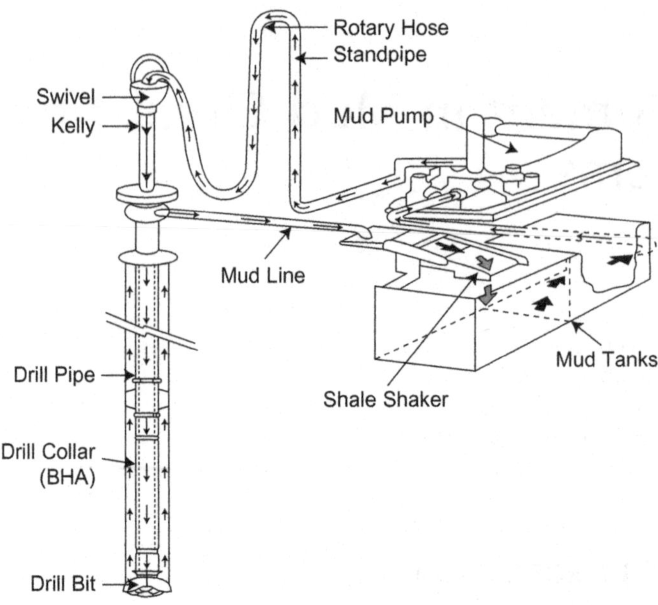

Figure 1.1 A typical mud circulating system.

Figure 1.2 A duplex pump. *(Courtesy of Great American.)*

of operation and maintenance, reliability, and the ability to operate over a wide range of pressures and flow rates by changing the diameters of the compression cylinders (pump liners) and pistons.

The two types of piston strokes are the single-action piston stroke and the double-action piston stroke. A pump that has double-action strokes in two cylinders is called a *duplex pump* (Figure 1.2). A pump that has single-action strokes in three cylinders is called a *triplex pump* (Figure 1.3). Triplex pumps are lighter and more compact than duplex pumps, their output

Figure 1.3 A triplex pump. *(Courtesy of TSC.)*

Figure 1.4 Pump liners. *(Courtesy of TSC.)*

pressure pulsations are not as great, and they are cheaper to operate. For these reasons, the majority of new pumps being placed into operation are of the triplex design. Normally, duplex pumps can handle higher flow rates, and triplex pumps can provide higher working pressure. However, for a given pump of fixed horsepower, the flow rate and working pressure can be adjusted by changing the sizes of the liners inside the pump. Some types of pump liners are shown in Figure 1.4. Changing the speed of the prime mover can also affect the mud flow rate in a certain range.

1.3 DRILL STRINGS

Figure 1.5 shows a drill string used in the oil and gas industry. The components of a drill string are described in this section according to the sequence in which they are run into the hole. A drill bit is installed at the bottom of the drill string. Three types of drill bits are used in the oil and gas industry: drag bits, cone bits (roller cutter bits), and PDC bits. Figure 1.6 shows two tricone bits, one with milled teeth and one with inserted teeth. Figure 1.7 shows a cutoff view of a tricone bit. Most bits are designed to install nozzles of different sizes. Drilling fluid passes through the bit nozzles at high velocity to clean the bit teeth and remove cuttings from the bottom of the hole. Figure 1.8 shows several types of bit nozzles, which are made of hard metal to resist erosion. Bit nozzle diameters often are expressed in 32nds of an inch. For example, if the bit nozzles are described as "12-13-13," this denotes that the bit contains one nozzle with a diameter of $^{12}/_{32}$ inch and two nozzles with diameters of $^{13}/_{32}$ inch. Odd-numbered nozzles are not available for nozzle sizes above 20.

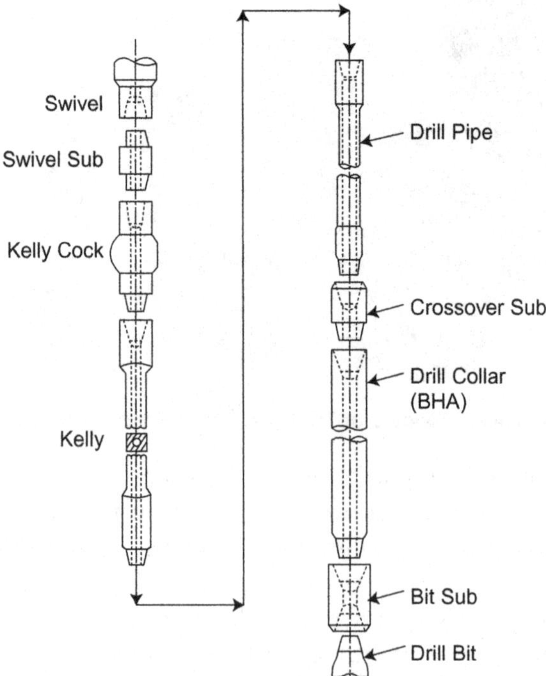

Figure 1.5 A drill string used in the petroleum industry.

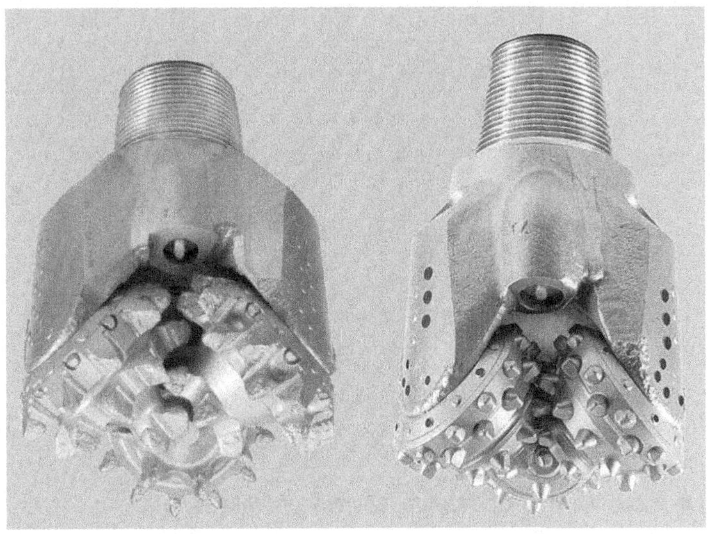

Figure 1.6 Typical tricone bits used in the petroleum industry. *(Courtesy of Lilin Petroleum Machinery Co., Ltd.)*

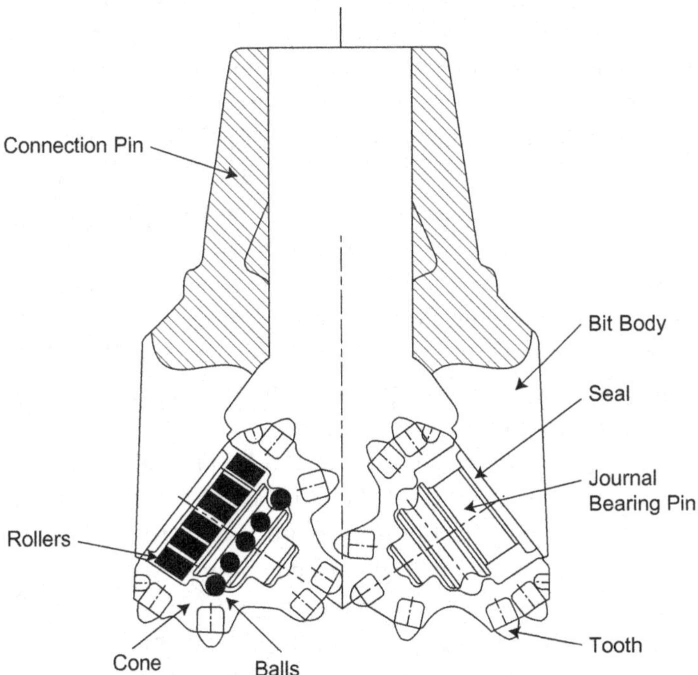

Figure 1.7 A cutoff view of a tricone bit.

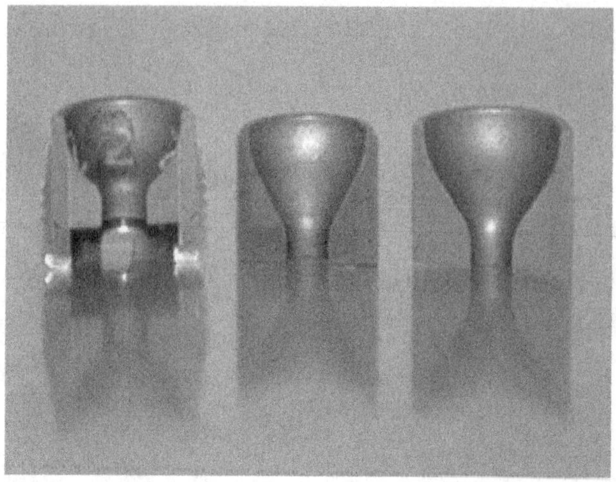

Figure 1.8 Cutoff views of bit nozzles. *(Courtesy of Schlumberger.)*

The drill bit is connected to drill collars through a bit sub, which is a short, thick wall pipe that has a threaded box on both ends. The bit sub is used for protecting the bottom threads of the bottom drill collar from wear due to the frequent drill bit connections. A drill collar is a thick wall pipe that gives weight to the drill bit, allowing the drill bit teeth to cut into the formation rock. Drill collars form a bottomhole assembly, with stabilizers installed at different distances from the bit (Figure 1.9) to control the well trajectory. The number of drill collars in a BHA depends on how much weight must be applied to the bit to make it advance efficiently (Bourgoyne et al., 1991). For directional and horizontal drilling, the BHA also includes mud motors and tools for measurements, such as measurements while drilling (MWD) and logging while drilling (LWD). The major components of a mud motor are shown in Figure 1.10. A pressure drop of a few hundred psi is required at the motor to generate torque for the rotating drill bit.

The drill pipe joints are above the BHA. The threads of the drill collar connections are usually not the same as the threads at the ends of the drill pipe joint. A special crossover sub is used to connect the drill pipe to the collars. The number of drill pipe joints used depends on the depth of the borehole. Conventional drill pipe joints are not designed to work in compression. After being pulled out of the hole for bit changes or well logging operations, the drill pipe string is broken into stand-by-stand sections and placed vertically between the drilling floor

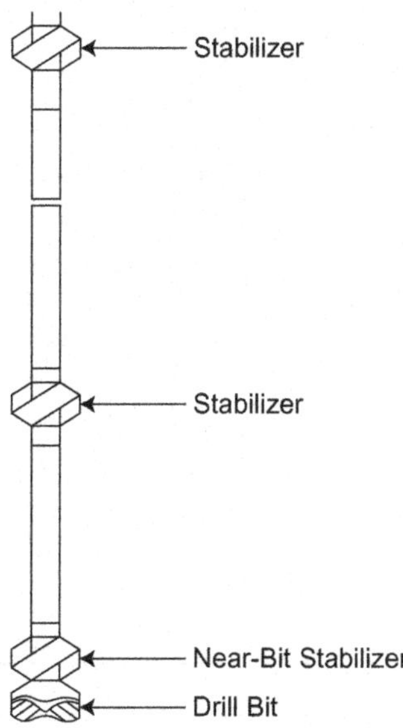

Figure 1.9 A bottomhole assembly used in the petroleum industry.

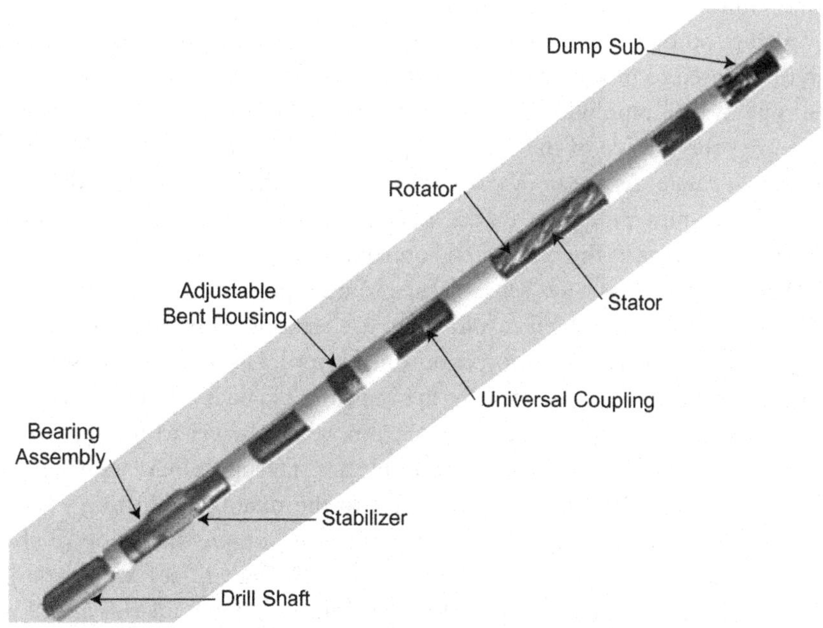

Figure 1.10 The major components of a mud motor. *(Courtesy of Lilin Petroleum Machinery Co., Ltd.)*

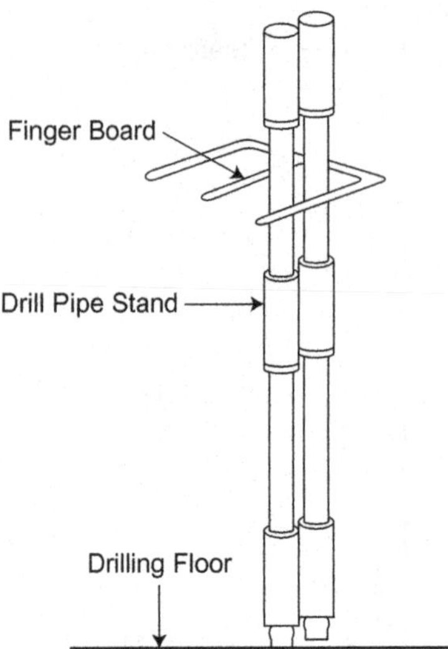

Figure 1.11 Drill pipe stands.

and the monkey board (Figure 1.11). The length of the drill pipe stands depends on the type of drill rig used. A stand usually contains two or three joints.

At the top of the drill string is the kelly cock sub. It is another crossover sub that is used to protect the bottom threads of the kelly. The kelly is a special type of drill pipe with a square or hexagonal cross-section. The rotary table grips the outside of the kelly and provides the torque to the drill string to make it rotate. As the borehole gets deeper, the drill pipe string is disconnected and a new pipe joint is added. The bottom thread of the kelly cock sub takes the wear of these repeated connections of drill pipe.

Above the kelly is a swivel sub that protects the swivel. Above the swivel sub is the swivel, which is split into two sections: a rotating section on the bottom and a nonrotating section on the top. The nonrotating section of the swivel is held in the mast by the traveling block and hoisting system. A sealed bearing allows the bottom section of the swivel to rotate, while the top section can be held in position by the traveling block. The swivel allows the drilling mud to flow through it to the rotating drill string.

For direct circulation, the drilling mud flows down the inside of the drill string to the drill bit, flows through the drill bit orifices (or nozzles), entrains the rock cuttings from the drill bit, and flows up the annulus between the drill string and the borehole. Once it reaches the surface, the mud is cleaned by the contaminant-removal equipment.

1.4 CONTAMINANT-REMOVAL EQUIPMENT

The contaminant-removal equipment is often called solid-removal equipment because it is mostly used for removing drilling cuttings in the mud returned from the borehole. A primary solid-removal equipment should include shale shakers, a degasser, a desander, a desilter, and a decanting centrifuge. All of these parts are installed on top of the mud tanks. As shown in Figure 1.12 (Moore, 1986), the order of operation is shale shakers, degasser, desander, desilter, and centrifuge.

The term *shale shaker* is used in mud drilling to cover all of the devices that in other industries might be differentiated as shaking screens, vibrating screens, and oscillating screens. All three of these types are used in the oil and gas industry, although most of them would probably fall into the vibrating screen classification. Figure 1.13 shows a shale shaker. Several factors affect the efficiency of shale shakers, including mud properties, screen mesh, vibrating frequency, and geometry of design. The particle-size separation made by a shale shaker screen is not simply that all particles larger than the stated screen mesh are

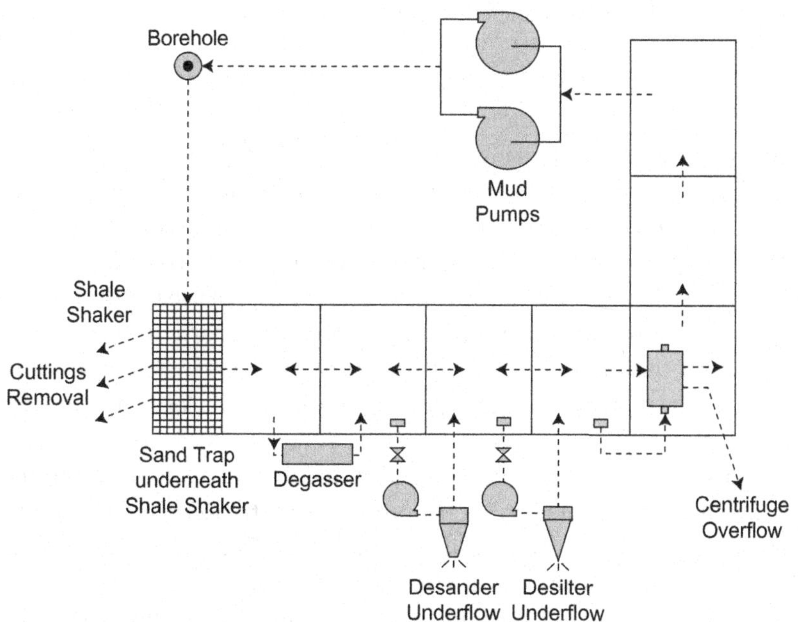

Figure 1.12 The solid-removal equipment.

Figure 1.13 A shale shaker *(Courtesy of TSC.)*

rejected and all smaller pieces pass through. The process of the screen moving at high vibratory speed prevents many undersized particles from passing through. Cuttings are not neat round balls, and no perfect method is available for measuring and stating their size. The median cut-size particle size is the size of half the particles that pass through and the size of the half that are rejected. The median particle size is much smaller than the mesh size for vibrating screens. Square mesh vibrating screens reject approximately 85 percent of the cuttings of the same size as the mesh.

Degassers are used to remove gas in the mud that is to be pumped into desanders using centrifugal pumps. The efficiency of the pumps drops significantly if gas exists in the mud. Degassers are essential to the solid-removal process involving viscous mud. Shale shakers can remove a good portion of the gas from badly gas-cut mud, especially if the yield point is lower than 10 lb/100 sq ft. A degasser is usually not necessary if the yield point of the mud is less than 6 lb/100 sq ft. If the degasser is a vacuum type (Figure 1.14) that requires power mud to operate the mud educator, the power mud should be taken from the degasser discharge compartment only. Violation of this rule will cause bypassing of the

Figure 1.14 A vacuum degasser. *(Courtesy of TSC.)*

desander or desilter when the degasser is in operation. Conventional gas-liquid separators (Figure 1.15) are also used in mud drilling.

Desanders (Figure 1.16), desilters (Figure 1.17), and the combination of them, called *mud cleaners* (Figure 1.18), are hydrocyclone-type equipment. If operated properly, hydrocyclones can perform the finest cut of any primary separation equipment operating on the full-flow circulating rate of an unweighted mud system. Understanding the most common designs of the hydrocyclone is essential to proper operation.

Figure 1.15 A gas-liquid separator. *(Courtesy of TSC.)*

Figure 1.16 A desander. *(Courtesy of TSC.)*

Figure 1.17 A desilter. *(Courtesy of TSC.)*

Figure 1.18 A mud cleaner. *(Courtesy of TSC.)*

The hydrocyclone (Figure 1.19) has a conical-shaped portion in which most of the settling takes place and a cylindrical feed chamber at the large end of the conical section. At the apex of the conical section is the underflow opening for the solids discharge. In operation, the underflow opening is usually at the bottom. Near the top end of the feed chamber, the inlet nozzle enters tangentially to the inside circumference and on a plane perpendicular to the top-to-bottom central axis of the hydrocyclone. The hydrocyclone obtains its centrifugal field from the tangential velocity of the slurry entering the feed chamber.

An axial velocity component is created by the axial thrust of the feed stream leaving the blind annular space of the feed chamber. The result is a downward-spiraling velocity labeled *S* in Figure 1.19, in which *T* and *A* represent the tangential and the axial velocity components, respectively. A hollow cylinder, called the vortex finder, extends axially

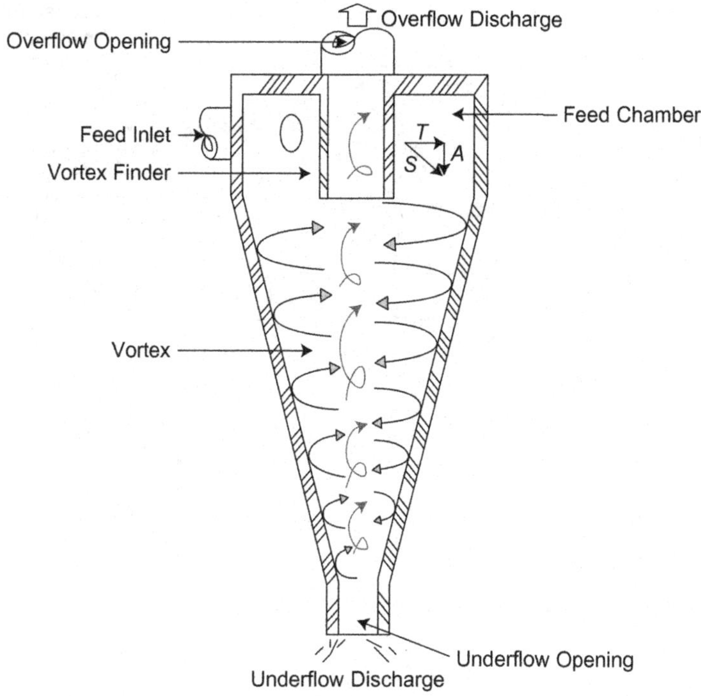

Figure 1.19 A hydrocyclone.

from the top into the barrel of the hydrocyclone past the inlet. The inside of the vortex finder forms the overflow outlet for the liquid discharge or effluent. The overflow opening is much larger than the underflow opening. The nominal size of a hydrocyclone is the largest inside diameter of the conical portion. All dimensions are critical to the operation of any specific design and size.

The decanting centrifuge (Figure 1.20) is a liquid–solid separation device used on drilling fluids that can remove (decant) all free liquid from separated solids particles, leaving only adsorbed moisture on the surface area. This adsorbed moisture does not contain soluble matter, such as chloride or colloidal suspended solids such as bentonite. The dissolved, suspended solids are associated with the continuous free liquid phase from which the decanting centrifuges the inert solids, and they remain with that liquid. The adsorbed moisture can be removed from the separated solids only by evaporation if necessary.

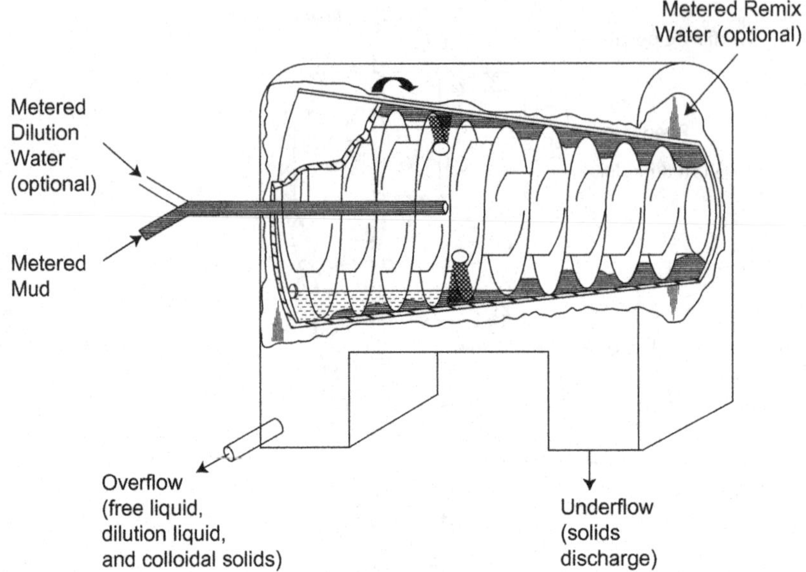

Figure 1.20 A decanting centrifuge.

SUMMARY

This chapter provided a brief introduction to the equipment in the mud circulating system. Mud pumps, drill pipes, drill collars, and bit nozzles are the essential components affecting mud drilling hydraulics and thus drilling performance, which is discussed in later chapters.

REFERENCES

Bourgoyne Jr. A.T., Millheim K.K., Chenevert, M.E., Young Jr. F.S., 1991. Applied Drilling Engineering. SPE Textbook Series.

Lyons, W.C., Guo, B., Graham, R.L., Hawley, G.D., 2009. Air and Gas Drilling Manual. Gulf Professional Publishing.

Moore, P.L., 1986. Drilling Practices Manual, second ed. PennWellBooks.

PROBLEMS

1.1 How can you adjust the mud flow rate from the mud pump?

1.2 What is the diameter in inches of a No. 16 nozzle?

1.3 What is the purpose of degassers in the mud circulating system?

CHAPTER TWO

Mud Hydraulics Fundamentals

Contents

2.1 INTRODUCTION

Mud hydraulics is considered one of the most important factors affecting mud drilling performance. The rate of penetration can be significantly increased using state-of-the-art techniques for hydraulics optimization to minimize drilling operation costs. The goal of the optimization is to make the maximum usage of a pump's power to help the bit to drill at maximum efficiency. This goal is achieved by minimizing the energy loss due to friction in the circulating system and using the saved energy to improve bit hydraulics. This chapter provides drilling engineers the essential fundamentals of mud hydraulics for pump selection (Chapter 3) and hydraulics optimization (Chapter 4).

2.2 CHARACTERIZATION OF DRILLING MUD

Different types of drilling mud are used in drilling operations based on their rheological behavior. This section describes the classification, rheology, and measurements of the properties of drilling mud used in the oil and gas industries.

2.2.1 Drilling Mud Classifications

Many different types of drilling mud are used in the industry. Their behaviors are very different in the drilling circulation systems. Pressure losses in flow conduits (drill string and annulus, in our case) are due to the resistance to flow. For a fluid particle flowing along the wall of a flow conduit, this resistance is the friction force from the wall. The friction force acts on the particle in the opposite direction of flow and tries to slow down the flow velocity of the particle. In the same way, the fluid particle along the wall exerts a friction force to the particle next to it. The friction forces, or flow resistances, between the fluid particle and the wall, and among fluid particles, depend on the fluid properties and the differences between the flow velocities of the particles. The study of this phenomenon of flow resistance is called rheology.

Rheology is basically the study of the flow or deformation of matter. It generally describes flow or deformation in terms of shear rate and shear stress. *Shear rate* is defined as the flow velocity gradient in the direction perpendicular to the flow direction. The higher the shear rate, the higher the friction between the flowing particles. The friction between particles is measured by the shear force per unit area of shearing layer, or *shear stress.*

Fluids are classified according to the different categories in rheological studies on the basis of their flow behaviors. Figure 2.1 shows five

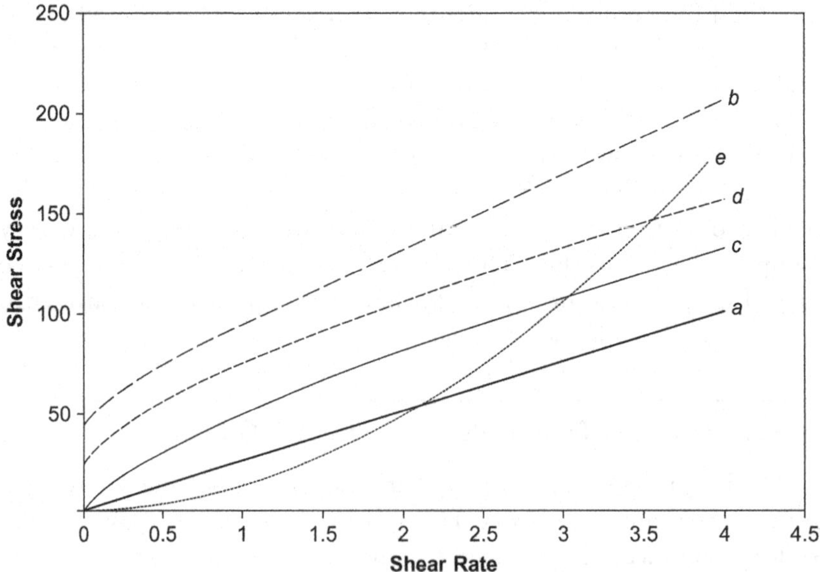

Figure 2.1 Different types of fluids found in drilling operations.

different types of fluids that are often encountered in the industry. Curve *a* characterizes the fluids that are the most common in nature. The shear stress is proportional to the shear rate, meaning that flow resistance increases linearly with flow deformation. Water and oil are examples of fluids in this category. These fluids are called Newtonian fluids.

Curve *b* shows a linear relationship between the shear rate and the shear stress, except in the low-shear-rate region. The shear stress takes a nonzero value at zero shear rate. This nonzero shear stress is called gel strength. It means that an initial force is required to deform and mobilize the fluid. Because of the plastic behavior, this type of fluid is called plastic fluid, or Bingham plastic fluid. Plastic fluids can be obtained by adding claylike solid particles to Newtonian fluids.

Curve *c* shows a nonlinear relationship between the shear rate and the shear stress. The flow resistance increases less than linearly with deformation. Fluid of this type is called pseudo plastic fluid, or Power Law fluid. Polymer solutions usually fall in this category.

Curve *d* shows a nonlinear relationship between the shear rate and the shear stress with a nonzero shear stress value at zero shear rate. An initial force is required to deform and mobilize the fluid. The flow resistance increases less than linearly with deformation. The behavior of this fluid was first modeled by Herschel and Bulkley (1926) and is called Herschel-Bulkley fluid.

Curve *e* also shows a nonlinear relationship between the shear rate and the shear stress. The flow resistance increases greater than linearly with deformation. This type of fluid is called dilatant fluid, which can be obtained by adding starchlike materials to Newtonian fluids.

2.2.2 Rheological Models

Different rheological models are used to describe the flow behavior of fluids. Newtonian fluids are described by the Newtonian model expressed as

$$\tau = \mu\dot{\gamma} \tag{2.1}$$

where

τ = shear stress, lb/100 ft^2 or Pa
μ = viscosity, cp or Pa-s
$\dot{\gamma}$ = shear rate, s^{-1}

The Bingham plastic model is used to describe the flow behavior of Bingham plastic fluids. The model is expressed as

$$\tau = \tau_y + \mu_p \dot{\gamma} \tag{2.2}$$

where

τ_y = yield point (YP), lb/100 ft^2 or Pa
μ_p = plastic viscosity (PV), cp or Pa-s

Apparently, the Bingham plastic model is a linear model that does not describe the flow behavior of the Bingham plastic fluid in the low-shear-rate region. The discrepancy is shown in Figure 2.2. The model parameter yield point (τ_y) overestimates the gel strength (τ_s) of fluid.

The behavior of both the pseudo plastic fluid and the dilatant fluid can be described by the so-called Power Law model expressed as

$$\tau = K\dot{\gamma}^n \tag{2.3}$$

where

K = consistency index, cp or Pa-s
n = flow behavior index, dimensionless

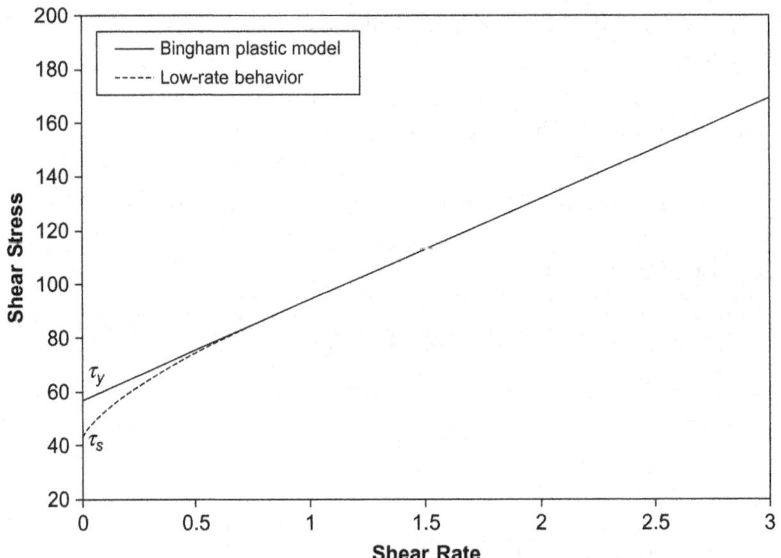

Figure 2.2 The Bingham plastic model does not describe the flow behavior of the Bingham plastic fluid in the low-shear-rate region.

When $n < 1$, the Power Law model describes the behavior of pseudo plastic fluids or Power Law fluids. When $n = 1$, the Power Law model describes the behavior of Newtonian fluids. When $n > 1$, the Power Law model describes the behavior of dilatant fluids.

The flow behavior of the Herschel-Bulkley fluids described by their model is expressed as

$$\tau = \tau_y + K\dot{\gamma}^n \tag{2.4}$$

Obviously, this three-parameter model is a general model that can be used to describe the behavior of all of the fluids shown in Figure 2.1.

Most drilling fluids are too complex to be characterized by the Newtonian model. Fluids that do not exhibit a direct proportionality between shear stress and shear rate are classified as *non-Newtonian*. Non-Newtonian fluids that are widely used in the drilling industry are the plastic and pseudo plastic fluids described by the Bingham plastic model and the Power Law model. The Herschel-Bulkley model is widely used by office engineers in designing fluid hydraulics.

These non-Newtonian fluids are *thixotropic* because the apparent viscosity (shear stress divided by shear rate) decreases with time after the shear rate is increased to a new value. This shear-thinning property is very desirable in drilling operations because we want low viscosity to reduce the circulating pressure in normal drilling operations, and we want high viscosity during circulation breaks to suspend drill cuttings in the annulus. At present, the thixotropic behavior of drilling fluids is not modeled mathematically. However, drilling fluids generally are stirred before measuring the apparent viscosities at various shear rates so steady-state conditions are obtained. Not accounting for thixotropy is satisfactory for most cases, but significant errors can occur when a large number of direction changes and dimension changes are present in the flow system.

In contrast to the plastic and pseudo plastic fluids, apparent viscosity of dilatant fluids increases with increasing shear rate. Since this shear-thickening property is not desirable in drilling operations, dilatant fluids are not purposely used as drilling fluids, but sometimes pseudo plastic fluids become dilatant fluids when significant amount of starchlike additives such as CMC are added to the system.

2.2.3 Measurements of Rheological Properties

Various types of instruments can be used to measure the rheological properties of fluids. The Fann 35 VG meter shown in Figure 2.3 is

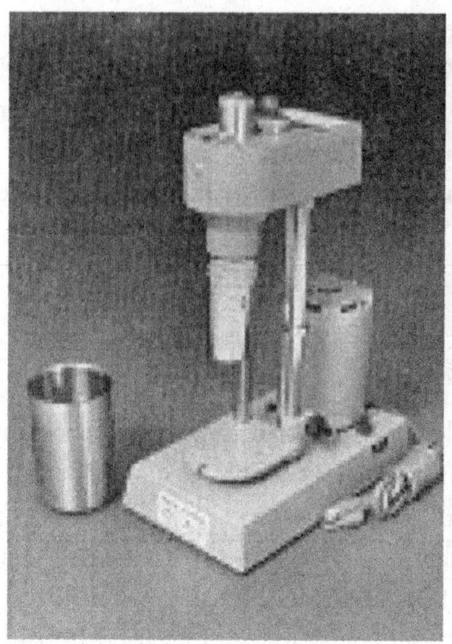
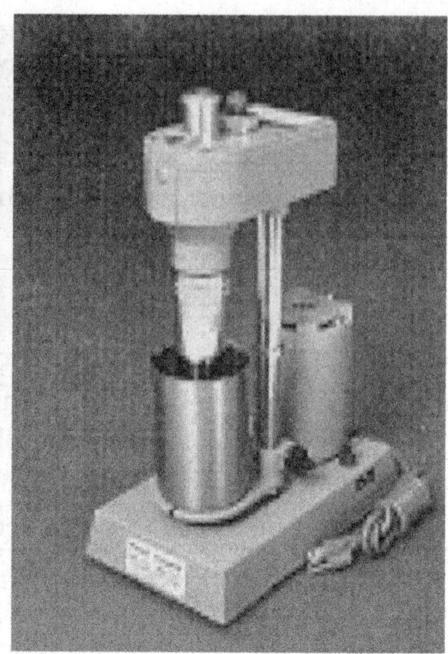

Figure 2.3 Fann VG meter 35A used for rheological property measurements. *(Images by author.)*

widely employed in the drilling industry. It can measure the rheological properties of all types of fluids quickly at six speeds. For a new fluid, however, it is important to first run the instrument at various speeds to get a complete set of shear rate and shear stress data. Plotting the data will help to identify the type of fluid. Then the rheological properties associated with the fluid model can be determined.

Newtonian Fluids

The viscosity of Newtonian fluids can be calculated using the following formula:

$$\mu = \frac{300}{N}\theta_N \tag{2.5}$$

where

N = rotary speed of Fann VG meter, rpm
θ_N = dial reading of Fann VG meter at rotary speed N

If the rotary speed of the Fann VG meter is set at 300 rpm, Eq. (2.5) degenerates to $\mu = \theta_{300}$, where θ_{300} is the dial reading of the Fann VG meter at the rotary speed of 300 rpm.

Bingham Plastic Fluids

The plastic viscosity (PV) of Bingham plastic fluids can be calculated using the following formula:

$$\mu_p = \frac{300}{N_2 - N_1}(\theta_{N_2} - \theta_{N_1}) \tag{2.6}$$

where

θ_{N_1} = dial reading of Fann VG meter at rotary speed N_1
θ_{N_2} = dial reading of Fann VG meter at rotary speed N_2

If the rotary speeds of the Fann VG meter are chosen to be $N_1 = 300$ rpm and $N_2 = 600$ rpm, Eq. (2.6) degenerates to $\mu_p = \theta_{600} - \theta_{300}$, where θ_{600} and θ_{300} are the dial readings of the Fann VG meter at the rotary speeds of 600 rpm and 300 rpm, respectively.

The yield point (YP) of Bingham plastic fluids can be calculated using the following formula:

$$\tau_y = \theta_{N_1} - \mu_p \frac{N_1}{300} \tag{2.7}$$

If the rotary speed of the Fann VG meter is chosen to be $N_1 = 300$ rpm, this equation degenerates to $\tau_y = \theta_{300} - \mu_p$.

Power Law Fluids

The flow behavior index of Power Law fluids can be calculated using the following formula:

$$n = \frac{\log\left(\dfrac{\theta_{N_2}}{\theta_{N_1}}\right)}{\log\left(\dfrac{N_2}{N_1}\right)} \tag{2.8}$$

If the rotary speeds of the Fann VG meter are chosen to be $N_1 = 300$ rpm and $N_2 = 600$ rpm, Eq. (2.8) degenerates to

$$n = 3.322 \log\left(\frac{\theta_{600}}{\theta_{300}}\right) \tag{2.9}$$

The consistency index of Power Law fluids can be calculated using the following formula:

$$K = \frac{510\,\theta_N}{(1.703N)^n} \tag{2.10}$$

If the rotary speed of the Fann VG meter is chosen to be $N = 300$ rpm, Eq. (2.10) degenerates to

$$K = \frac{510\,\theta_{300}}{(511)^n} \tag{2.11}$$

Herschel-Bulkley Fluids

The fluid yield stress τ_y is normally taken as the 3 rpm reading, with the flow behavior index n and the consistency index K then calculated from the 600 or 300 rpm values or graphically. The approximate yield stress τ_y, commonly known as the low-shear-rate yield point, should be determined by

$$\tau_y = 2\,\theta_3 - \theta_6 \tag{2.12}$$

The fluid flow index n is given by

$$n = 3.322 \log\left(\frac{\theta_{600} - \tau_y}{\theta_{300} - \tau_y}\right) \tag{2.13}$$

The fluid consistency index K is calculated by

$$K = 500\frac{(\theta_{300} - \tau_y)}{(511)^n} \tag{2.14}$$

For water-based drilling fluids containing large amounts of viscous polymers and thus high θ_{600} values, Eq. (2.12) can yield overstated values of τ_y.

At high shear rates, it is acceptable to treat Herschel-Bulkley fluids as Power Law fluids. The assumption is that the log-log slope of the Herschel-Bulkley flow equation is numerically close to that of the Power Law flow equation.

2.3 HYDRAULICS MODELS

The flow behavior of drilling mud can be described using mathematical models called hydraulics models. These models define the relationship between flow rate and pressure drop for a given geometry of flow conduit and fluid properties. The relationship also depends on flow regime.

2.3.1 Flow Regimes

The regimes most commonly encountered in drilling are *laminar*, *turbulent*, and *transitional*. In a laminar flow, the fluid behaves like a series of parallel layers moving at uniform or near-uniform velocity. There is no large-scale movement of fluid particles between layers. The fluid layers nearest the center of the pipe or annulus generally move faster than the layers adjacent to the pipe wall or wellbore. Turbulent flow is characterized by velocity fluctuations among the fluid stream particles, both parallel and axial to the mean flow stream. These fluctuations break down the boundaries between the fluid layers, resulting in a chaotic flow pattern.

Transitional flow exhibits characteristics of both laminar and turbulent regimes. It describes the often hard-to-define region where flow is neither completely laminar nor completely turbulent. Also reported in the literature is an additional fluid regime called *plug flow*. It describes the low-velocity, sublaminar condition of a fluid moving as a homogeneous, relatively undisturbed body. This flow regime has not been found in normal drilling conditions.

It is usually preferred to see laminar flow in the annulus to move cuttings up the hole and to prevent erosion. Turbulent flow, on the other hand, is more desirable at the bottom of the hole because it promotes cleaning and cuttings removal. While they are conceptually easy to visualize, flow regimes may be difficult to identify. Not only does fluid behavior vary within the circulating system, but more than one flow regime may exist at the same point in the system. For example, while the main flow stream in the annulus may exhibit laminar behavior, the adjacent fluid at the pipe boundary may be in turbulent flow.

Newtonian Fluids

The most common method for determining a fluid's flow regime is by calculating its Reynolds number. For Newtonian fluids inside pipe, the Reynolds number is defined as

$$N_{\mathrm{Re}} = \frac{\rho v d}{\mu} \tag{2.15}$$

where

ρ = fluid density, kg/m^3
d = inside diameter of pipe, m
μ = fluid viscosity, Pa-s

and

$$v = \frac{q}{0.7854d^2} \tag{2.16}$$

where

v = average flow velocity, m/s
q = flow rate, m^3/s

For annular flow, the Reynolds number becomes

$$N_{Re} = 0.816 \times \frac{\rho v(d_2 - d_1)}{\mu} \tag{2.17}$$

where

d_2 = hole or casing diameter, m
d_1 = outside diameter of pipe, m

and

$$v = \frac{q}{0.7854(d_2^2 - d_1^2)} \tag{2.18}$$

The term $0.816(d_2 - d_1)$ in Eq. (2.17) is the equivalent circular diameter of a slot representation of the annulus (Bourgoyne et al., 1986).

In U.S. field units (where ρ is given in ppg; v in ft/s; q in gpm; d, d_1, and d_2 in inches; and μ in cp), Eqs. (2.15) through (2.18), respectively, become

$$N_{Re} = 928 \times \frac{\rho v d}{\mu} \tag{2.19}$$

$$v = \frac{q}{2.448d^2} \tag{2.20}$$

$$N_{Re} = 757 \times \frac{\rho v(d_2 - d_1)}{\mu} \tag{2.21}$$

$$v = \frac{q}{2.448(d_2{}^2 - d_1{}^2)} \tag{2.22}$$

As a general guideline, Reynolds numbers of less than 2,100 indicate laminar flow, while Reynolds numbers greater than 4,000 indicate turbulent flow. Between these values, flow is considered transitional.

Illustrative Example 2.1

A 10.5-ppg Newtonian fluid with a viscosity of 30 cp is circulating at 250 gpm in an 8¾-in-diameter wellbore. Determine the flow regime inside a 4½-in OD, 16.60-lb/ft drill pipe (3.826-in ID), and in the drill pipe/hole annulus.

Solution

Inside the drill pipe:

$$v = \frac{250}{2.448(3.826)^2} = 6.98\,\text{ft/s}$$

$$N_{\text{Re}} = 928\frac{(10.5)(6.98)(3.826)}{30} = 8{,}674$$

Since $N_{\text{Re}} > 4{,}000$, turbulent flow exists inside the drill pipe.

In the annulus:

$$v = \frac{250}{2.448(8.75^2 - 4.5^2)} = 1.82\,\text{ft/s}$$

$$N_{\text{Re}} = 757\frac{(10.5)(1.82)(8.75 - 4.5)}{30} = 2{,}038$$

Since $N_{\text{Re}} < 2{,}100$, laminar flow exists in the annular space.

Unfortunately, determining flow regimes is seldom this straightforward. Laminar flow has been observed under controlled conditions for Reynolds numbers as low as 1,200 and as high as 40,000 (Bourgoyne et al., 1986), although we do not usually encounter such extremes in drilling operations.

Bingham Plastic Fluids

For Bingham plastic fluids, the equations for the Newtonian fluids need to be modified by defining an apparent viscosity to account for the plastic viscosity and yield point. For pipe flow, the definition is

$$\mu_a = \mu_p + \frac{6.66\tau_y d}{v} \tag{2.23}$$

For the annular flow, the definition is

$$\mu_a = \mu_p + \frac{5\tau_y(d_2 - d_1)}{v} \tag{2.24}$$

Equations (2.23) and (2.24) are valid for U.S. field units. When expressed in SI units, the constant 6.66 becomes 0.1669, and the constant 5 becomes 0.1253.

Thus, for Bingham plastic fluids, Eqs. (2.15), (2.17), (2.19), and (2.21), respectively, become

$$N_{Re} = \frac{\rho v d}{\mu_a} \tag{2.25}$$

$$N_{Re} = 0.816 \times \frac{\rho v (d_2 - d_1)}{\mu_a} \tag{2.26}$$

$$N_{Re} = 928 \times \frac{\rho v d}{\mu_a} \tag{2.27}$$

$$N_{Re} = 757 \times \frac{\rho v (d_2 - d_1)}{\mu_a} \tag{2.28}$$

Illustrative Example 2.2

A 10.5-ppg Bingham plastic fluid with a plastic viscosity of 20 cp and yield point of 5 lb/100 ft^2 is circulating at 250 gpm in an 8¾-inch-diameter wellbore. Determine the flow regime inside a 4½-in OD, 16.60-lb/ft drill pipe (3.826-in ID), and in the drill pipe/hole annulus.

Solution

Inside the drill pipe:

$$v = \frac{250}{2.448 \times (3.826)^2} = 6.98 \text{ ft/s}$$

$$\mu_a = 20 + \frac{6.66 \times 5 \times 3.826}{6.98} = 38.25 \text{ cp}$$

$$N_{Re} = 928 \times \frac{10.5 \times 6.98 \times 3.826}{38.25} = 6{,}803$$

Since $N_{Re} > 4{,}000$, turbulent flow exists inside the drill pipe.

In the annulus:

$$v = \frac{250}{2.448 \times (8.75^2 - 4.5^2)} = 1.81 \text{ ft/s}$$

$$\mu_a = 20 + \frac{5 \times 5 \times (8.75 - 4.5)}{1.81} = 78.70 \text{ cp}$$

$$N_{Re} = 757 \times \frac{10.5 \times 1.81 \times (8.75 - 4.5)}{78.70} = 777$$

Since $N_{Re} < 2{,}100$, laminar flow exists in the annular space.

Using these equations, the criterion for turbulent flow is the same as for Newtonian fluids, with laminar flow occurring below a Reynolds number of 2,100.

Power Law Fluids

The concept of apparent viscosity can also be used for Power Law fluids for Reynolds number calculations. Equations (2.23) and (2.24), respectively, become

$$\mu_a = \frac{Kd^{(1-n)}}{96v^{(1-n)}}\left(\frac{3+1/n}{0.0416}\right)^n \tag{2.29}$$

$$\mu_a = \frac{K(d_2-d_1)^{1-n}}{144v^{(1-n)}}\left(\frac{2+1/n}{0.0208}\right)^n \tag{2.30}$$

If Dodge and Metzner's (1959) correlation is used, the Reynolds number for pipe flow and annular flow can be respectively expressed in U.S. field units as

$$N_{\text{Re}} = 89{,}100\frac{\rho v^{2-n}}{K}\left(\frac{0.0416d}{3+1/n}\right)^n \tag{2.31}$$

and

$$N_{\text{Re}} = 109{,}000\frac{\rho v^{2-n}}{K}\left[\frac{0.0208(d_2-d_1)}{2+1/n}\right]^n \tag{2.32}$$

When expressed in SI units, these two equations become

$$N_{\text{Re}} = 743.5\frac{\rho(3.281v)^{2-n}}{K}\left(\frac{1.638d}{3+1/n}\right)^n \tag{2.33}$$

and

$$N_{\text{Re}} = 909.5\frac{\rho(3.281v)^{2-n}}{K}\left[\frac{0.819(d_2-d_1)}{2+1/n}\right]^n \tag{2.34}$$

The turbulence criterion for Power Law fluids is based on a critical Reynolds number ($N_{\text{Re}c}$) that depends on the value of the flow behavior index. A simple correlation for estimating the critical Reynolds number at the upper limit of laminar flow is

$$N_{\text{Re}c} = 3{,}470 - 1370n \tag{2.35}$$

Illustrative Example 2.3

A 10.5-ppg Power Law fluid with a consistency index of 20 cp equivalent and flow behavior index of 0.8 is circulating at 250 gpm in an 8-¾-in-diameter wellbore. Determine the flow regime inside a 4½-in OD, 16.60-lb/ft drill pipe (3.826-in ID), and in the drill pipe/hole annulus.

Solution

Inside the drill pipe:

$$v = \frac{250}{2.448 \times (3.826)^2} = 6.98 \text{ ft/s}$$

$$N_{\text{Re}} = 89{,}100 \times \frac{10.5 \times 6.98^{(2-0.8)}}{20} \times \left[\frac{0.0416 \times 3.826}{3 + 1/0.8}\right]^{0.8} = 34{,}788$$

$$N_{\text{Rec}} - L_{am} = 3{,}470 - 1{,}370 \times 0.8 = 2{,}374$$
$$N_{\text{Rec}} - T_{ur} = 4{,}270 - 1{,}370 \times 0.8 = 3{,}174$$

Since $N_{\text{Re}} > 3{,}174$, turbulent flow exists inside the drill pipe.

In the annulus:

$$v = \frac{250}{2.448 \times (8.75^2 - 4.5^2)} = 1.81 \text{ ft/s}$$

$$N_{\text{Re}} = 109{,}000 \times \frac{10.5 \times 1.81^{(2-0.8)}}{20} \times \left[\frac{0.0208 \times (8.75 - 4.5)}{2 + 1/0.8}\right]^{0.8} = 6{,}523$$

Since $N_{\text{Re}} > 3{,}174$, turbulent flow exists in the annular space.

For the region between transitional and turbulent flow, the critical Reynolds number is

$$N_{\text{Rec}} = 4{,}270 - 1370n \tag{2.36}$$

Herschel-Bulkley Fluids

For Herschel-Bulkley fluids, the Reynolds number can be calculated using the following equations in U.S. field units.

Inside the drill pipe:

$$N_{\text{Re}} = \frac{2(3n+1)}{n}\left[\frac{\rho v^{(2-n)}\left(\frac{d}{2}\right)^n}{\tau_y\left(\frac{d}{2v}\right)^n + K\left(\frac{3n+1}{nC_c}\right)^n}\right] \tag{2.37}$$

In the annulus:

$$N_{Re} = \frac{4(2n+1)}{n}\left[\frac{\rho v^{(2-n)}\left(\frac{d_2-d_1}{2}\right)^n}{\tau_y\left(\frac{d_2-d_1}{2v}\right)^n + K\left(\frac{2(2n+1)}{nC_a^*}\right)^n}\right] \tag{2.38}$$

where the constants C_c and C_a^* are expressed respectively as follows:

$$C_c = 1 - \left(\frac{1}{2n+1}\right)\frac{\tau_y}{\tau_y + K\left[\frac{(3n+1)q}{n\pi(d/2)^3}\right]^n} \tag{2.39}$$

and

$$C_a^* = 1 - \left(\frac{1}{n+1}\right)\frac{\tau_y}{\tau_y + K\left\{\frac{2q(2n+1)}{n\pi[(d_2/2)-(d_1/2)][(d_2/2)^2-(d_1/2)^2]}\right\}^n} \tag{2.40}$$

The critical Reynolds number N_{Rec} inside the drill pipe and in the annulus can be estimated respectively as

$$N_{Rec} = \left[\frac{4(3n+1)}{ny}\right]^{\frac{1}{1-z}} \tag{2.41}$$

and

$$N_{Rec} = \left[\frac{8(2n+1)}{ny}\right]^{\frac{1}{1-z}} \tag{2.42}$$

where

$$y = \frac{\log(n) + 3.93}{50} \tag{2.43}$$

$$z = \frac{1.75 - \log(n)}{7} \tag{2.44}$$

The critical Reynolds number (N_{Rec}) is the criterion for determining the flow regime. The flow becomes turbulent once it is over the critical Reynolds number; otherwise it is laminar flow.

Illustrative Example 2.4

A 10.5-ppg Herschel-Bulkley fluid with a consistency index of 20 cp equivalent, a flow behavior index of 0.8, and a yield stress of 5 lb/100 ft^2 is circulating at 250 gpm in an 8¾-in-diameter wellbore. Determine the flow regime inside a 4½-in OD, 16.60-lb/ft drill pipe (3.826-in ID) and in the drill pipe/hole annulus.

Solution

Inside the drill pipe:

$$v = \frac{250}{2.448(3.826)^2} = 6.98\ \text{ft/s}$$

$$C_c = 1 - \left(\frac{1}{2\times 0.8+1}\right)\frac{5}{5+0.04177\left[\dfrac{(3\times 0.8+1)0.557}{0.8\times \pi(0.319/2)^3}\right]^{0.8}} = 0.7513$$

$$N_{\text{Re}} = \frac{2(3\times 0.8+1)}{0.8}\left[\frac{10.5\times 7.48\times 6.98^{(2-0.8)}\left(\dfrac{0.319}{2}\right)^{0.8}}{5\left(\dfrac{0.319}{2\times 6.98}\right)^{0.8}+0.04177\left(\dfrac{3\times 0.8+1}{0.8\times 0.7513}\right)^{0.8}}\right] = 11{,}975$$

The critical Reynolds number N_{Rec} inside the drill pipe is calculated as follows:

$$y = \frac{\log(0.8)+3.93}{50} = 0.0767$$

$$z = \frac{1.75-\log(0.8)}{7} = 0.2638$$

$$N_{\text{Rec}} = \left[\frac{4(3\times 0.8+1)}{0.8\times 0.0767}\right]^{\frac{1}{1-0.2638}} = 1{,}537$$

Since $N_{\text{Re}} > 1{,}537$, turbulent flow exists inside the drill pipe.

In the annulus:

$$v = \frac{250}{2.448(8.75^2-4.5^2)} = 1.81\ \text{ft/s}$$

$$C_a^* = 1 - \left(\frac{1}{0.8+1}\right)\frac{5}{5+0.04177\left\{\dfrac{2\times 0.557(2\times 0.8+1)}{0.8\times \pi[(0.729/2)-(0.375/2)][(0.729/2)^2-(0.375/2)^2]}\right\}^{0.8}}$$

$$C_a^* = 0.552$$

$$N_{\text{Re}} = \frac{4(2\times 0.8+1)}{0.8}\left[\frac{10.5\times 7.48\times 1.81^{(2-0.8)}\left(\dfrac{0.729-0.375}{2}\right)^{0.8}}{5\left(\dfrac{0.729-0.375}{2\times 1.81}\right)^{0.8}+0.04177\left(\dfrac{2(2\times 0.8+1)}{0.8\times 0.552}\right)^{0.8}}\right] = 1{,}506$$

The critical Reynolds number N_{Rec} in the annulus is calculated as follows:

$$N_{Rec} = \left[\frac{8(2 \times 0.8 + 1)}{0.8 \times 0.0767}\right]^{\frac{1}{1-0.2638}} = 2{,}737$$

Since $N_{Re} < 2{,}737$, laminar flow exists in the annular space.

2.3.2 Pressure Loss

For drilling fluid to flow through the circulating system, it must overcome frictional forces between the fluid layers, solid particles, pipe wall, and borehole wall. The pump pressure corresponds to the sum of these forces:

$$p_p = \Delta p_s + \Delta p_{dp} + \Delta p_{dc} + \Delta p_{mt} + \Delta p_b + \Delta p_{dca} + \Delta p_{dpa} \tag{2.45}$$

where

p_p = pump pressure, psi or kPa
Δp_s = pressure loss in the surface equipment, psi or kPa
Δp_{dp} = pressure loss inside drill pipe, psi or kPa
Δp_{dc} = pressure loss inside drill collar, psi or kPa
Δp_{mt} = pressure drop inside mud motor, psi or kPa
Δp_b = pressure drop at bit, psi or kPa
Δp_{dca} = pressure loss in the drill collar annulus, psi or kPa
Δp_{dpa} = pressure loss in the drill pipe annulus, psi or kPa

If the total frictional pressure loss to and from the bit is called the *parasitic pressure loss* Δp_d, then

$$\Delta p_d = \Delta p_s + \Delta p_{dp} + \Delta p_{dc} + \Delta p_{dca} + \Delta p_{dpa} \tag{2.46}$$

and (if no mud motor used)

$$p_p = \Delta p_b + \Delta p_d \tag{2.47}$$

For a given fluid type, flow regime, and type of conduit (in pipe or annulus), all the components of the parasitic pressure loss can be calculated with the same pressure loss equation.

The surface equipment consists of a standpipe, a rotary hose, a swivel, and a kelly pipe. Table 2.1 shows the inner diameter and length of each for some typical combinations. In field applications, the total pressure loss

Table 2.1 Inner Equipment Diameter and Length for Typical Combinations

	Combination 1				Combination 2				Combination 3				Combination 4			
	ID		Length		ID		Length		ID		Length		ID		Length	
Component	in	cm	ft	m	in	cm	ft	m	in	cm	ft	m	in	cm	ft	m
Standpipe	3	7.6	40	12.2	3.5	8.9	40	12.2	4	10.2	45	13.7	4	10.2	45	13.7
Rotary hose	2	5.1	45	13.7	2.5	6.4	55	16.8	3	7.6	55	16.8	3	7.6	55	16.8
Swivel	2	5.1	4	1.2	2.5	6.4	5	1.5	2.5	6.4	5	1.5	3	7.6	6	1.8
Kelly pipe	2.25	5.7	40	12.2	3.3	8.3	40	12.2	3.3	8.3	40	12.2	4	10.2	40	12.2

Table 2.2 The Equivalent Drill Pipe Length for Typical Equipment Combinations

Equivalent Drill Pipe	Combination 1		Combination 2		Combination 3		Combination 4	
	ft	m	ft	m	ft	m	ft	m
3.5″, 13.3 lb/ft	437	133	161	49				
4.5″, 16.6 lb/ft			761	232	479	146	340	104
5″, 19.5 lb/ft					816	249	576	176

in the surface equipment is not calculated based on the geometry of each piece of equipment. Instead, the pressure loss is estimated using an equivalent length of drill pipe. Table 2.2 presents the equivalent drill pipe length data for the typical combinations.

The general procedure for calculating system pressure losses is as follows:

1. Determine the fluid velocity (or Reynolds number) at the point of interest.
2. Calculate the critical velocity (or Reynolds number) to determine whether the fluid is in laminar or turbulent flow.
3. Choose the appropriate pressure loss equation based on the rheological model and flow regime applied to the point of interest.

In field applications, calculate both the actual Reynolds number N_{Re} and the critical Reynolds number $N_{\mathrm{Re}c}$. If $N_{\mathrm{Re}} > N_{\mathrm{Re}c}$, the flow is turbulent, while if $N_{\mathrm{Re}} < N_{\mathrm{Re}c}$, it is laminar. If the actual and critical Reynolds numbers are approximately equal, then perform pressure loss calculations for both flow regimes and use the results that give the larger pressure loss.

Pressure loss in a conduit depends on the type of fluid. Different flow equations have been used in the industry to calculate pressure losses in drill strings and annuli. Based on the Fanning equation (Bourgoyne et al., 1986), the gradient of frictional pressure drop in a conduit is expressed as follows:

$$\frac{dp_f}{dL} = \frac{f\rho_f\bar{v}^2}{25.8d} \tag{2.48}$$

where

p_f = frictional pressure, psi or kPa
L = pipe length, ft or m
f = Fanning friction factor, dimensionless
$\bar{v}$ = average velocity, ft/s or m/s
d = equivalent pipe inner diameter, in or m

The constant 25.8 in U.S. units is 519 in the SI units.

The Fanning friction factor f is a function of the Reynolds number N_{Re} and a term called the relative roughness. The *relative roughness* is defined as the ratio of the absolute roughness over the pipe diameter, where the absolute roughness represents the average depth of pipe-wall irregularities. Table 2.3 shows the absolute roughness of some pipe surfaces.

For laminar flow, the friction factor is replaced by $f = \frac{16}{N_{\mathrm{Re}}}$. Several empirical correlations for the determination of friction factor for fully developed turbulent flow in circular pipe have been presented, including those by Colebrook (1938):

$$\frac{1}{\sqrt{f}} = -4\log\left(0.269\delta/d + \frac{1.255}{N_{\mathrm{Re}}\sqrt{f}}\right) \tag{2.49}$$

where the Reynolds number is defined as

$$N_{\mathrm{Re}} = \frac{928\rho_f \bar{v} d}{\mu} \tag{2.50}$$

The constant 928 in U.S. units is 1 in SI units.

The δ is the absolute roughness of the pipe surface in inches (in) or meter (m). The selection of an appropriate absolute roughness for a given application is often difficult. Fortunately, in rotary drilling applications involving the use of relatively viscous drilling fluids, for most wellbore geometries the relative roughness is usually less than 0.0004 in all sections. For these conditions, the friction factor for smooth pipe can be applied for most engineering calculations:

$$\frac{1}{\sqrt{f}} = 4\log(N_{\mathrm{Re}}\sqrt{f}) - 0.395 \tag{2.51}$$

Table 2.3 Absolute Roughness of Some Pipe Surfaces

	Absolute Roughness	
Type of Pipe	**in**	**mm**
Riveted steel	0.00025 ~ 0.0025	0.00635 ~ 0.0635
Concrete	0.000083 ~ 0.00083	0.0021 ~ 0.021
Cast iron	0.000071	0.0018034
Galvanized iron	0.000042	0.0010668
Asphalted cast iron	0.000033	0.0008382
Commercial steel	0.000013	0.0003302
Drawn tubing	0.0000004	0.00001016

For smooth pipe, Colebrook's (1938) friction factor function can be simplified to

$$f = \frac{0.0791}{N_{\mathrm{Re}}^{0.25}} \tag{2.52}$$

which was first presented by Blasius (1913).

Chen's (1979) correlation has an explicit form and gives similar accuracy to the Colebrook-White equation (Gregory and Fogarasi, 1985) that was employed for generating the friction factor chart widely used in the petroleum industry. Chen's correlation takes the following form:

$$f = \left(-4\log\left\{\frac{\varepsilon}{3.7065} - \frac{5.0452}{N_{\mathrm{Re}}}\log\left[\frac{\varepsilon^{1.1098}}{2.8257} + \left(\frac{7.149}{N_{\mathrm{Re}}}\right)^{0.8981}\right]\right\}\right)^{-2} \tag{2.53}$$

where the relative roughness is defined as $\varepsilon = \frac{\delta}{d}$.

Newtonian Fluids

If the friction factor in Eq. (2.48) is replaced by $f = \frac{16}{N_{\mathrm{Re}}}$, the pressure loss under laminar flow inside the drill string and in the annulus can be estimated using the following equations respectively:

$$\Delta p_f = \frac{\mu v}{1{,}500 d^2}\Delta L \tag{2.54}$$

$$\Delta p_f = \frac{\mu v}{1{,}000(d_2 - d_1)^2}\Delta L \tag{2.55}$$

where

Δp_f = pressure loss, psi or kPa
ΔL = length of conduit, ft or m

These two equations are valid in U.S. field units. When expressed in SI units, the constant 1,500 becomes 0.0313 and 1,000 becomes 0.0209.

Using the Chen friction factor correlation allows for accurate prediction of frictional pressure loss in turbulent flow. However, in many cases, using the Blasius correlation gives a result that is accurate enough for frictional pressure calculations. Substituting Eq. (2.52) into Eq. (2.48) and rearranging the latter yield the pressure loss expressions for inside the drill string and in the annulus as follows respectively:

$$\Delta p_f = \frac{\rho^{0.75} v^{1.75} \mu^{0.25}}{1{,}800 d^{1.25}}\Delta L \tag{2.56}$$

and

$$\Delta p_f = \frac{\rho^{0.75} v^{1.75} \mu^{0.25}}{1{,}396(d_2 - d_1)^{1.25}} \Delta L \tag{2.57}$$

These two equations are valid in U.S. field units. When expressed in SI units, the constant 1,800 becomes 631.8 and the constant 1,396 becomes 490.

Bingham Plastic Fluids

For Bingham plastic fluids, the pressure loss under laminar flow inside the drill string and in the annulus can be estimated using the following equations respectively:

$$\Delta p_f = \left(\frac{\mu_p v}{1{,}500 d^2} + \frac{\tau_y}{225 d} \right) \Delta L \tag{2.58}$$

and

$$\Delta p_f = \left[\frac{\mu_p v}{1{,}000(d_2 - d_1)^2} + \frac{\tau_y}{200(d_2 - d_1)} \right] \Delta L \tag{2.59}$$

These two equations are valid in U.S. field units. When expressed in SI units, the constants 1,500, 225, 1,000, and 200 become 31.33, 187.5, 20.88, and 166.7, respectively.

The pressure loss under turbulent flow inside the drill string and in the annulus can be estimated using the following equations respectively:

$$\Delta p_f = \frac{\rho^{0.75} v^{1.75} \mu_p^{0.25}}{1{,}800 d^{1.25}} \Delta L \tag{2.60}$$

and

$$\Delta p_f = \frac{\rho^{0.75} v^{1.75} \mu_p^{0.25}}{1{,}396(d_2 - d_1)^{1.25}} \Delta L \tag{2.61}$$

These two equations are valid in U.S. field units. When expressed in SI units, the constant 1,800 becomes 6,320, and 1,396 becomes 4,901.

Illustrative Example 2.5

Determine the system pressure loss for the following well:

Total depth: 9,950 ft (3,036 m)
Casing: 9⅝ in, 43.5 lb/ft (8.755-in ID), set at 6,500 ft (1,982 m)
Open hole: 8½ in from 6,500 ft to 9,950 ft
Drill pipe: 9,500 ft of 4½ in, 16.6 lb/ft (3.826-in ID)
Drill collar: 450 ft of 6¾-in OD and 2¼-in ID
Surface equipment: Combination 3
Mud weight: 10.5 ppg
Plastic viscosity: 35 cp
Yield point: 6 lb/100 ft^2
Mud flow rate: 300 gpm

Solution

According to Table 2.2 shown earlier, the pressure loss through the surface equipment is equivalent to that through 479 ft of 4½ in, 16.6-lb/ft drill pipe.

Inside the drill pipe:

$$v = \frac{300}{2.448(3.826)^2} = 8.37\,\text{ft/s}$$

$$\mu_a = 35 + \frac{6.66(6)(3.826)}{8.37} = 53\,\text{cp}$$

$$N_{\text{Re}} = 928\frac{(10.5)(8.37)(3.826)}{53} = 5{,}887 > 2{,}100, \text{turbulent flow}$$

$$\Delta p_f = \frac{(10.5)^{0.75}(8.37)^{1.75}(35)^{0.25}}{1{,}800(3.826)^{1.25}}(9{,}500 + 479) = 605\,\text{psi}$$

Inside the drill collar:

$$v = \frac{300}{2.448(2.25)^2} = 24.2\,\text{ft/s}$$

$$\mu_a = 35 + \frac{6.66(6)(2.25)}{24.2} = 39\,\text{cp}$$

$$N_{\text{Re}} = 928\frac{(10.5)(24.2)(2.25)}{39} = 13{,}604 > 2{,}100, \text{turbulent flow}$$

$$\Delta p_f = \frac{(10.5)^{0.75}(24.2)^{1.75}(35)^{0.25}}{1{,}800(2.25)^{1.25}}(450) = 340\,\text{psi}$$

(Continued)

Illustrative Example 2.5 *(Continued)*

In the cased-hole annulus:

$$v = \frac{300}{2.448(8.755^2 - 4.5^2)} = 2.17\,\text{ft/s}$$

$$\mu_a = 35 + \frac{5(6)(8.755 - 4.5)}{2.17} = 94\,\text{cp}$$

$$N_{Re} = 757\frac{(10.5)(2.17)(8.755 - 4.5)}{94} = 781 < 2{,}100, \text{ laminar flow}$$

$$\Delta p_f = \left[\frac{(35)(2.17)}{1{,}000(8.755 - 4.5)^2} + \frac{6}{200(8.755 - 4.5)}\right](6500) = 73\,\text{psi}$$

In the open-hole/drill pipe annulus:

$$v = \frac{300}{2.448(8.5^2 - 4.5^2)} = 2.36\,\text{ft/s}$$

$$\mu_a = 35 + \frac{5(6)(8.5 - 4.5)}{2.36} = 86\,\text{cp}$$

$$N_{Re} = 757\frac{(10.5)(2.36)(8.5 - 4.5)}{86} = 872 < 2{,}100, \text{ laminar flow}$$

$$\Delta p_f = \left[\frac{(35)(2.36)}{1{,}000(8.5 - 4.5)^2} + \frac{6}{200(8.5 - 4.5)}\right](3{,}000) = 38\,\text{psi}$$

In the open-hole/drill collar annulus:

$$v = \frac{300}{2.448(8.5^2 - 6.75^2)} = 4.59\,\text{ft/s}$$

$$\mu_a = 35 + \frac{5(6)(8.5 - 6.75)}{4.59} = 46\,\text{cp}$$

$$N_{Re} = 757\frac{(10.5)(4.59)(8.5 - 6.75)}{46} = 1{,}388 < 2{,}100, \text{ laminar flow}$$

$$\Delta p_f = \left[\frac{(35)(4.59)}{1{,}000(8.5 - 6.75)^2} + \frac{6}{200(8.5 - 6.75)}\right](450) = 31\,\text{psi}$$

The total system pressure loss is

$$\Delta p_d = 605 + 340 + 73 + 38 + 31 = 1{,}087\,\text{psi} = 7{,}394\,\text{kPa}$$

Power Law Fluids

For Power Law fluids, the pressure loss under laminar flow inside the drill string and in the annulus can be estimated using the following equations respectively:

$$\Delta p_f = \left[\left(\frac{96v}{d}\right)\left(\frac{3n+1}{4n}\right)\right]^n \frac{K}{300d}\Delta L \tag{2.62}$$

and

$$\Delta p_f = \left[\left(\frac{144v}{d_2-d_1}\right)\left(\frac{2n+1}{3n}\right)\right]^n \frac{K}{300(d_2-d_1)}\Delta L \tag{2.63}$$

Equations (2.62) and (2.63) are valid in U.S. field units. When expressed in SI units, they take the following form:

$$\Delta p_f = 0.0019152\left[\left(\frac{8v}{d}\right)\left(\frac{3n+1}{4n}\right)\right]^n \frac{K}{d}\Delta L \tag{2.64}$$

and

$$\Delta p_f = 0.0019152\left[\left(\frac{12v}{d_2-d_1}\right)\left(\frac{2n+1}{3n}\right)\right]^n \frac{K}{(d_2-d_1)}\Delta L \tag{2.65}$$

There is no simple correlation found to estimate the friction factor for pressure loss under turbulent flow of Power Law fluids. Therefore, the original form of the friction loss equation has to be used. The following equations are employed for pipe flow and annular flow respectively:

$$\Delta p_f = \frac{f\rho v^2}{25.8d}\Delta L \tag{2.66}$$

and

$$\Delta p_f = \frac{f\rho v^2}{21.1(d_2-d_1)}\Delta L \tag{2.67}$$

Illustrative Example 2.6

Determine the system pressure loss for the well in Illustrative Example 2.5 assuming Power Law fluids with a consistency index of 20 cp equivalent and a blow behavior index of 0.8.

Solution

According to Table 2.2, the pressure loss through the surface equipment is equivalent to that through 479 ft of 4½-in, 16.6-lb/ft drill pipe.

(Continued)

Illustrative Example 2.6 *(Continued)*

Inside the drill pipe:

$$v = \frac{300}{2.448 \times (3.826)^2} = 8.37\,\text{ft/s}$$

$$N_{Re} = 89{,}100 \times \frac{10.5 \times 8.37^{(2-0.8)}}{20} \times \left[\frac{0.0416 \times 3.826}{3 + 1/0.8}\right]^{0.8} = 43{,}258$$

$$N_{Rec}\text{Lam} = 4{,}470 - 1{,}370 \times 0.8 = 2{,}374$$

$$N_{Rec}\text{Tur} = 4{,}270 - 1{,}370 \times 0.8 = 3{,}174$$

Since $N_{Re} > 3{,}174$, turbulent flow exists inside the drill pipe.

$$f = \frac{0.0791}{43{,}258^{0.25}} = 0.005485$$

$$\Delta p_f = \frac{0.005485 \times 10.5 \times 8.37^2}{25.8 \times 3.826} \times [9{,}500 + 479] = 408\,\text{psi}$$

Inside the drill collar:

$$v = \frac{300}{2.448 \times (2.25)^2} = 24.2\,\text{ft/s}$$

$$N_{Re} = 89{,}100 \times \frac{10.5 \times 24.2^{(2-0.8)}}{20} \times \left[\frac{0.0416 \times 2.25}{3 + 1/0.8}\right]^{0.8} = 101{,}140$$

Since $N_{Re} > 3{,}174$, turbulent flow exists inside the drill collar.

$$f = \frac{0.0791}{101{,}140^{0.25}} = 0.00444$$

$$\Delta p_f = \frac{0.00444 \times 10.5 \times 24.2^2}{25.8 \times (2.25)} \times 450 = 211\,\text{psi}$$

In the cased-hole annulus:

$$v = \frac{300}{2.448 \times (8.755^2 - 4.5^2)} = 2.17\,\text{ft/s}$$

$$N_{Re} = 109{,}000 \times \frac{10.5 \times 2.17^{(2-0.8)}}{20} \times \left[\frac{0.0208 \times (8.755 - 4.5)}{2 + 1/0.8}\right]^{0.8}$$

$$= 8{,}117 > 3{,}174 \text{ turbulent flow}$$

$$f = \frac{0.0791}{8{,}117^{0.25}} = 0.00833$$

$$\Delta p_f = \frac{0.00833 \times 10.5 \times 2.17^2}{21.1 \times (8.755 - 4.5)} \times 6{,}500 = 30\,\text{psi}$$

In the open-hole/drill pipe annulus:

$$v = \frac{300}{2.448 \times (8.5^2 - 4.5^2)} = 2.36\,\text{ft/s}$$

$$N_{\text{Re}} = 109{,}000 \times \frac{10.5 \times 2.36^{(2-0.8)}}{20} \times \left[\frac{0.0208 \times (8.5 - 4.5)}{2 + 1/0.8}\right]^{0.8}$$

$$= 8{,}544 > 3{,}174 \text{ turbulent flow}$$

$$f = \frac{0.0791}{8{,}544^{0.25}} = 0.00823$$

$$\Delta p_f = \frac{0.00823 \times 10.5 \times 2.36^2}{21.1 \times (8.5 - 4.5)} \times 3{,}000 = 17\,\text{psi}$$

In the open-hole/drill collar annulus:

$$v = \frac{300}{2.448 \times (8.5^2 - 6.75^2)} = 4.59\,\text{ft/s}$$

$$N_{\text{Re}} = 109{,}000 \times \frac{10.5 \times 4.59^{(2-0.8)}}{20} \times \left[\frac{0.0208 \times (8.5 - 6.75)}{2 + 1/0.8}\right]^{0.8}$$

$$= 9{,}798 > 3{,}174 \text{ turbulent flow}$$

$$f = \frac{0.0791}{9{,}798^{0.25}} = 0.00795$$

$$\Delta p_f = \frac{0.00795 \times 10.5 \times 4.59^2}{21.1 \times (8.5 - 6.75)} \times 450 = 21\,\text{psi}$$

The total system pressure loss is

$$\Delta p_d = 408 + 211 + 30 + 17 + 21 = 687\,\text{psi} = 4{,}737\,\text{kPa}$$

Herschel-Bulkley Fluids

For Herschel-Bulkley fluids, the pressure loss under laminar flow can be calculated in U.S. field units using the following equations.

Inside the drill pipe:

$$\Delta p_f = \frac{4K}{14400d}\left\{\left(\frac{\tau_y}{K}\right) + \left[\left(\frac{3n+1}{nC_c}\right)\left(\frac{8q}{\pi d^3}\right)\right]^n\right\}\Delta L \tag{2.68}$$

In the annulus:

$$\Delta p_f = \frac{4K}{14400(d_2 - d_1)}\left\{\left(\frac{\tau_y}{K}\right) + \left[\left(\frac{16(2n+1)}{n \times C_a^*(d_2 - d_1)}\right)\left(\frac{q}{\pi(d_2^2 - d_1^2)}\right)\right]^n\right\}\Delta L \tag{2.69}$$

For turbulent flow, the pressure loss inside the drill pipe and in the annulus are estimated respectively as

$$\Delta p_f = \frac{f_c q^2 \rho}{1421.22 d^5} \Delta L \quad (2.70)$$

and

$$\Delta p_f = \frac{f_a q^2 \rho}{1421.22(d_2 - d_1)(d_2^2 - d_1^2)^2} \Delta L \quad (2.71)$$

The friction factors f_c inside the drill pipe and f_a in the annulus are calculated respectively as

$$f_c = \gamma (C_c N_{Re})^{-z} \quad (2.72)$$

and

$$f_a = \gamma (C_a^* N_{Re})^{-z} \quad (2.73)$$

Illustrative Example 2.7

Determine the system pressure loss for the following well:

Total depth: 9,950 ft (3,036 m)
Casing: 9⅝ in, 43.5 lb/ft (8.755-in ID), set at 6,500 ft (1,982 m)
Open hole: 8½ in from 6,500 ft to 9,950 ft
Drill pipe: 9,500 ft of 4½ in, 16.6 lb/ft (3.826-in ID)
Drill collar: 450 ft of 6¾-in OD and 2¼-in ID
Surface equipment: Combination 3
Mud weight: 10.5 ppg
Yield point: 6 lb/100 ft^2
Consistency index: 20 eq. cp
Flow behavior index: 0.8
Mud flow rate: 300 gpm

Solution

According to Table 2.2 shown earlier, the pressure loss through the surface equipment is equivalent to that through 479 ft of 4½ in, 16.6-lb/ft drill pipe.

Inside the drill pipe:

$$v = \frac{300}{2.448(3.826)^2} = 8.37\,\text{ft/s}$$

Like the calculation in Illustrative Example 2.4, the values of the Reynolds number N_{Re}, the critical Reynolds number N_{Rec}, y, z, and C_c are calculated following:

$$N_{Re} = 14{,}563,\ N_{Rec} = 1{,}537,\ N_{Re} > N_{Rec}, \text{turbulent flow}$$

$$C_c = 0.7481,\ y = 0.0767,\ z = 0.2638$$

$$f_c = 0.0767(0.7481\times 14{,}563)^{-0.2638} = 0.0066$$

$$\Delta p_f = \frac{0.0066\times 0.6684^2\times 10.5\times 7.48}{1421.22\times (3.826/12)^5}(9{,}500+479) = 491\ \text{psi}$$

Inside the drill collar:

$$v = \frac{300}{2.448(2.25)^2} = 24.2\ \text{ft/s}$$

$$N_{Re} = 65{,}119,\ N_{Rec} = 1{,}537,\ N_{Re} > N_{Rec}, \text{turbulent flow}$$

$$C_c = 0.8666,\ y = 0.0767,\ z = 0.2638$$

$$f_c = 0.0767(0.8666\times 65{,}119)^{-0.2638} = 0.0043$$

$$\Delta p_f = \frac{0.0043\times 0.6684^2\times 10.5\times 7.48}{1421.22\times (2.25/12)^5}(450) = 204\ \text{psi}$$

In the cased-hole annulus:

$$v = \frac{300}{2.448(8.755^2 - 4.5^2)} = 2.17\ \text{ft/s}$$

$$N_{Re} = 1{,}818,\ N_{Rec} = 2{,}737,\ N_{Re} < N_{Rec}, \text{laminar flow}$$

$$C_a^* = 0.5487$$

$$\Delta p_f = \frac{4\times 0.04177}{14400(0.73-0.375)}\times\left\{\left(\frac{6}{0.04177}\right) + \left[\left(\frac{16(2\times 0.8+1)}{0.8\times 0.5487(0.73-0.375)}\right)\left(\frac{0.6684}{\pi(0.73^2-0.375^2)}\right)\right]^{0.8}\right\}(6{,}500) = 42\ \text{psi}$$

In the open-hole/drill pipe annulus:

$$v = \frac{300}{2.448(8.5^2-4.5^2)} = 2.36\ \text{ft/s}$$

$$N_{Re} = 2{,}080,\ N_{Rec} = 2{,}737,\ N_{Re} < N_{Rec}, \text{laminar flow}$$

$$C_a^* = 0.5589$$

$$\Delta p_f = \frac{4\times 0.04177}{14400(0.708-0.375)}\times\left\{\left(\frac{6}{0.04177}\right) + \left[\left(\frac{16(2\times 0.8+1)}{0.8\times 0.5589(0.708-0.375)}\right)\left(\frac{0.6684}{\pi(0.708^2-0.375^2)}\right)\right]^{0.8}\right\}(3{,}000) = 21\ \text{psi}$$

(Continued)

Illustrative Example 2.7 *(Continued)*
In the open-hole/drill collar annulus:

$$v = \frac{300}{2.448(8.5^2 - 6.75^2)} = 4.59\,\text{ft/s}$$

$$N_{\text{Re}} = 5{,}218,\ N_{\text{Rec}} = 2{,}737,\ N_{\text{Re}} < N_{\text{Rec}},\ \text{turbulent flow}$$

$$C_a^* = 0.5589$$

$$f_a = 0.0767(0.5589 \times 5{,}218)^{-0.2638} = 0.008$$

$$\Delta p_f = \frac{0.008 \times 0.6684^2 \times 10.5 \times 7.48}{1421.22 \times (0.708 - 0.563) \times (0.708^2 - 0.563^2)^2}(450) = 18\,\text{psi}$$

The total system pressure loss is

$$\Delta p_d = 491 + 204 + 42 + 21 + 18 = 776\,\text{psi} = 5{,}279\,\text{kPa}$$

The Generalized Pressure Loss Model

The total parasitic pressure loss in a drilling circulation system includes the frictional pressure loss in the surface equipment Δp_s, frictional pressure losses in the drill pipe Δp_{dp} and drill collars Δp_{dc}, and frictional pressure losses in the drill collar annulus Δp_{dca} and the drill pipe annulus Δp_{dpa}. If each term of the parasitic pressure loss is computed for the usual case of turbulent flow, examining the equations for turbulent flow yields

$$\Delta p_d = cq^m \tag{2.74}$$

where m is a constant that theoretically has a value near 1.75 for turbulent flow, and c is a constant that depends on the mud properties and wellbore geometry. Considering that laminar flow may exist in some annular sections, the constant m may take a value less than 1.75.

The values of c and m can be estimated by matching the calculated pressure losses with the model $\Delta p_d = cq^m$ at two flow rates. At a given depth of interest, suppose pressure losses at flow rates q_1 and q_2 are calculated as Δp_{d1} and Δp_{d2}, respectively. The values of c and m in the flow rate range can be determined by

$$m = \frac{\log\left(\dfrac{\Delta p_{d2}}{\Delta p_{d1}}\right)}{\log\left(\dfrac{q_2}{q_1}\right)} \tag{2.75}$$

$$c = \frac{\Delta p_{d1}}{q_1{}^m} \tag{2.76}$$

or

$$c = \frac{\Delta p_{d2}}{q_2{}^m} \tag{2.77}$$

Spreadsheet *c & m-Values.xls* attached to this book can be used for estimating the c and m values for Bingham plastic fluids. In field operations, the values of c and m can be determined with the following procedure:

1. At the current hole depth, before tripping out to change the bit, circulate the drilling fluid at two flow rates (q_1, q_2) with the bit off bottom and record pump pressures (p_{p1}, p_{p2}) at the two corresponding rates. The flow rates should be selected to reflect the range of flow rate to be used while drilling with the next bit.
2. When the bit is pulled out to surface, connect the bit to the kelly directly, circulate the drilling fluid at the same two flow rates (q_1, q_2), and record two pump pressures. These two pump pressures approximate pressure drops (Δp_{b1}, Δp_{b2}) across the bit nozzles at the given flow rates.
3. Determine the parasitic pressures at the two flow rates by

$$\Delta p_{d1} = p_{p1} - \Delta p_{b1} \tag{2.78}$$

$$\Delta p_{d2} = p_{p2} - \Delta p_{b2} \tag{2.79}$$

Determine the values of c and m in the flow rate range by Eqs. (2.75) and (2.76).

2.3.3 Surge and Swab Pressure

When a drill string is run in a hole, it forces drilling fluid up the annulus and out of the flow line. At the same time, the mud immediately adjacent to the pipe is dragged downhole. The resulting piston effect generates a surge pressure that is added to the hydrostatic pressure. Excessive surge pressures can increase the wellbore pressure to such a degree that it can lose circulation. Conversely, when a drill string is pulled out of a hole, fluid flows down the annulus to fill the resulting void. This causes a suction effect, generating a swab pressure that can lower the differential pressure and possibly bring formation fluid into the borehole.

Calculating surge and swab pressures can be a complex undertaking, depending on the pipe configuration and the hole geometry. Burkhardt (1961) developed a relationship between well geometry and the effect of the fluid being dragged by the pipe. Based on Burkhardt's work, the effective annular velocity is equal to

$$v_e = v_m - \kappa v_p \tag{2.80}$$

where

v_e = effective annular velocity, ft/s or m/s
v_m = mud velocity, ft/s or m/s
v_p = pipe velocity, ft/s or m/s

and κ is referred to as the clinging constant, which is a function of annular geometry. Burkhardt presented a chart for determining the value of κ in both laminar flow and turbulent flow. We found that the chart can be replaced by the following correlations with minimal error. For laminar flow, the correlation is

$$\kappa = 0.275\left(\frac{d_p}{d_h}\right) + 0.25 \tag{2.81}$$

where

d_p = outer diameter of pipe, in or mm
d_h = hole diameter, in or mm

For turbulent flow, the correlation is

$$\kappa = 0.1\left(\frac{d_p}{d_h}\right) + 0.41 \tag{2.82}$$

For closed-end pipes, such as a casing string with a float shoe, the mud velocity can be calculated by

$$v_m = -v_p\left(\frac{d_p^2}{d_h^2 - d_p^2}\right) \tag{2.83}$$

For open-end pipes, the mud velocity can be calculated by

$$v_m = -v_p\left(\frac{4d_p^2(d_h - d_p)^2 - 3d_p^4}{4d_p^2(d_h - d_p)^2(d_h^2 - d_p^2) + 6d_p^4}\right) \tag{2.84}$$

Illustrative Example 2.8

Calculate the surge pressure generated by a 10¾-in casing string under the following conditions, and predict whether the total borehole pressure will exceed the formation fracture gradient. Assume that the casing is effectively "closed" with a float shoe and laminar flow in the annulus.

Casing depth: 6,400 ft (1,951 m) TVD
Fracture gradient: 0.82 psi/ft
Hole diameter: 14¾ in
Mud weight: 15.5 ppg
Plastic viscosity: 37 cp
Yield point: 6 lb/100 ft^2
Pipe velocity: −110 ft/min (the negative sign denotes downward velocity)

Solution

$$v_m = -\frac{-110}{60}\left(\frac{10.75^2}{14.75^2 - 10.75^2}\right) = 2.08\,\text{ft/s}$$

$$\kappa = 0.275\left(\frac{10.75}{14.75}\right) + 0.25 = 0.45$$

$$v_e = 2.08 - 0.45\left(\frac{110}{60}\right) = 1.25\,\text{ft/s}$$

Use the annular flow pressure loss equation for the laminar flow:

$$\Delta p_f = \left[\frac{(37)(1.265)}{1{,}000(14.75 - 10.75)^2} + \frac{6}{200(14.75 - 10.75)}\right](6{,}400) = 67\,\text{psi}$$

The equivalent mud weight (EMW) is

$$\text{EMW} = 15.5 + \frac{67}{0.052(6{,}400)} = 15.7\,\text{ppg}$$

The equivalent mud weight of the fracture gradient (EMW$_f$) is

$$\text{EMW}_f = \frac{0.82}{0.052} = 15.8\,\text{ppg} > 15.7\,\text{ppg}$$

Therefore, the borehole will be safe during running the casing at this depth.

Illustrative Example 2.9

Using the Bingham plastic model, calculate the swab pressure generated by a 10-¾-in casing string under the following conditions, and predict whether the

(Continued)

Illustrative Example 2.9 *(Continued)*
total borehole pressure will be lower than the formation pore gradient. Assume that the casing is fully opened and laminar flow in the annulus.

Casing depth: 6,400 ft (1,951 m)
Pore gradient: 0.78 psi/ft
Hole diameter: 14¾ in
Mud weight: 15.5 ppg
Plastic viscosity: 37 cp
Yield point: 6 lb/100 ft^2
Pipe velocity: 110 ft/min

Solution

$$v_m = -\frac{110}{60}\times\left[\frac{4\times10.75^2\times(14.75-10.75)^2-3\times10.75^4}{4\times10.75^2\times(14.75-10.75)^2\times(14.75^2-10.75^2)+6\times10.75^4}\right]$$

$$=0.0718\,\text{ft/s}$$

$$k=0.275\times\frac{10.75}{14.75}+0.25=0.45$$

$$v_e=0.0718+0.45\times\frac{110}{60}=0.90\,\text{ft/s}$$

Use the annular flow pressure loss equation for the laminar flow:

$$\Delta p_f=\left[\frac{37\times0.90}{1{,}000\times(14.75-10.75)^2}+\frac{6}{200\times(14.75-10.75)}\right]\times6{,}400=61\,\text{psi}$$

$$\text{EMW}=15.5-\frac{61}{0.052\times6{,}400}=15.32\,\text{ppg}$$

The equivalent mud weight of the pore gradient (EMW_P) is

$$\text{EMW}_P=\frac{0.78}{0.052}=15\,\text{ppg}<15.32\,\text{ppg}$$

Therefore, the borehole will be safe while running the casing at this depth.

2.3.4 Pressure Drop at the Bit

The purpose of installing jet nozzles on a bit is to improve the cleaning action of the drilling fluid at the bottom of the hole. Because of the small diameter of bit nozzles, fluids reach high velocities inside the nozzle. This velocity is called nozzle velocity and is expressed as

$$v_n=0.32086\frac{q}{A_T} \tag{2.85}$$

where

v_n = nozzle velocity, ft/s or m/s
q = mud flow rate, gpm or m^3/s
A_T = total nozzle area, in^2 or m^2

The constant 0.32086 becomes 1 in SI units.

Drill bit pressure losses do not result primarily from friction forces but are due to the acceleration of the drilling fluid through the bit nozzles. The bit pressure drop is expressed as

$$\Delta p_b = \frac{\rho q^2}{12{,}031 C_d^2 A_T^2} \tag{2.86}$$

where

C_d = nozzle discharge coefficient, dimensionless

In the SI system, where Δp_b is expressed in kPa, the 12,031 becomes 2,000. The discharge coefficient accounts for the nonideal conditions, including the viscous frictional effects. Its value has been determined experimentally for bit nozzles by several researchers. They indicated that the discharge coefficient may be as high as 0.98, depending on the nozzle type and size, but they recommended a value of 0.95 as a more practical limit. Since the viscous frictional effects are essentially negligible for flow in short nozzles, Eq. (2.86) is valid for both Newtonian and non-Newtonian liquids.

An expression of nozzle velocity as a function of pressure drop can be derived from Eq. (2.86) as

$$v_n = C_d \sqrt{\frac{\Delta p_b}{8.074 \times 10^{-4} \rho}} \tag{2.87}$$

The hydraulic power of drilling fluid at the bit is one of the indicators of the hole-cleaning capacity of the fluid. It is expressed as

$$P_{Hb} = \frac{\Delta p_b q}{1{,}714} \tag{2.88}$$

where

P_{Hb} = bit hydraulic power, hp or w

The constant 1,714 becomes 1 in SI units.

The hydraulic impact force is another indicator of the hole-cleaning capacity of the drilling fluid. It is expressed as

$$F_j = 0.01823 C_d q \sqrt{\rho \Delta p_b} \tag{2.89}$$

where

F_j = bit hydraulic impact force, lb_f or N

The constant 0.01823 becomes 1.4142 in SI units.

2.3.5 The Cuttings-Carrying Capacity of Mud

A minimum mud flow rate is required for carrying drill cuttings to the surface. This minimum flow rate can be estimated based on the minimum required mud velocity, which should be higher than the drill cuttings slip velocity. Unfortunately, because of the complex geometry and boundary conditions involved, analytical expressions describing drill cuttings slip velocity have been obtained for only very idealized conditions.

For a drill cutting falling in a Newtonian fluid, its terminal slip velocity can be expressed as

$$v_{sl} = 1.89 \sqrt{\frac{d_s}{f_p}\left(\frac{\rho_s - 7.48\rho_f}{7.48\rho_f}\right)} \tag{2.90}$$

where

v_{sl} = cuttings slip velocity, ft/s or m/s
d_s = equivalent cuttings diameter, in or m
ρ_s = cuttings density, lb/ft^3 or kg/m^3
ρ_f = fluid density, ppg or kg/m^3
f_p = particle friction factor, dimensionless

The constants 1.89 and 7.48 in U.S. units are 2.97 and 1 in SI units, respectively.

The equivalent cuttings diameter depends on several factors, including formation lithology, bit type, rate of penetration, and rotary speed at the bit. It can be estimated on the basis of data from offset drilling. For a given bit to drill a given formation of rock, the cuttings size can be reduced by using a low rate of penetration and a high rotary speed. The following formula gives an approximation of equivalent cutting diameter:

$$d_s = 0.2 \frac{ROP}{RPM} \tag{2.91}$$

where

ROP = rate of penetration, ft/hr or m/hr
RPM = rotary speed of bit, rpm

The constant 0.2 in U.S. units is 0.0167 in SI units.

The particle friction factor f is a function of the Reynolds number N_{Re} and particle sphericity ψ. The sphericity is defined as the surface area of a sphere containing the same volume as the particle divided by the surface area of the particle. A conservative value for cuttings sphericity is 0.8. Engineering charts are available for finding the values of the friction factor (Bourgoyne et al., 1986). Fang et al. (2008) developed the following correlation to replace the charts:

$$f_p = 10^\wedge\left(A' + B'\log(N_{ReP}) + C'[\log(N_{ReP})]^2\right) \tag{2.92}$$

where

$$A' = 2.2954 - 2.2626\,\psi + 4.4395\,\psi^2 - 2.9825\,\psi^3 \tag{2.93}$$

$$B' = -0.4193 - 1.9014\,\psi + 3.3416\,\psi^2 - 2.0409\,\psi^3 \tag{2.94}$$

$$C' = 0.1117 + 0.0553\,\psi - 0.1468\,\psi^2 + 0.1145\,\psi^3 \tag{2.95}$$

where the particle Reynolds number is defined as

$$N_{ReP} = \frac{928\rho_f v_{sl} d_s}{\mu} \tag{2.96}$$

where

μ = viscosity of Newtonian fluid, cp or Pa-s

The constant 928 in U.S. units is 1 in SI units.

Because the slip velocity is implicitly involved in Eqs. (2.90) and (2.92), the slip velocity can only be solved numerically (trial and error). A computer program called *Cuttings Slip Velocity.xls* is attached to this book for easy calculations. To calculate the particle slip velocity using Table 2.4, (1) select a unit system, (2) update the data in the Input Data column, and (3) click the Solution button and obtain the result.

In non-Newtonian fluids, an analytical solution for cuttings terminal slip velocity has not been developed. For Bingham plastic fluids, there is a critical (minimum) cuttings diameter for it to slip (Bourgoyne et al.,

Table 2.4 Computer Program *Cuttings Slip Velocity.xls*

Input Data	**U.S. Units**	**SI Units**
Particle diameter	0.25 in	
Particle spherity	0.8	
Drilling fluid viscosity	6 cp	
Drilling fluid density	12 ppg	
Cuttings specific gravity	2.7	
Solution		
$A' = 2.2954 - 2.2626\psi + 4.4395\psi^2 - 2.9825\psi^3$	= 1.7996	
$B' = -0.4193 - 1.9014\psi + 3.3416\psi^2 - 2.0409\psi^3$	= −0.8467	
$C' = 0.1117 + 0.0553\psi - 0.1468\psi^2 + 0.1145\psi^3$	= 0.1206	
$N_{ReP} = \dfrac{928\rho_f v_{sl} d_s}{\mu}$	= 240	
$f_p = 10\wedge\left(A' + B'\log(N_{ReP}) + C'[\log(N_{ReP})]^2\right)$	= 2.9350	
$v_{sl} = 1.89\sqrt{\dfrac{d_s}{f_p}\left(\dfrac{\rho_s - 7.48\rho_f}{7.48\rho_f}\right)}$	= 0.5166 ft/s	= 0.157 m/s

1986). This critical diameter is directly proportional to the gel strength of the fluid. Multiple correlations have been developed to estimate the cuttings slip velocity in non-Newtonian fluids. These correlations are documented by Chien (1971), Moore (1986), and Walker and Mayes (1975). However, Eq. (2.90) gives conservative estimates for the cuttings terminal slip velocity in non-Newtonian fluids.

The minimum required mud velocity should be higher than the drill cuttings slip velocity by an additional amount called *transport velocity*—that is,

$$v_{min} = v_{sl} + v_{tr} \tag{2.97}$$

where

v_{min} = minimum required mud velocity, ft/s or m/s
v_{tr} = transport velocity, ft/s or m/s

The required transport velocity depends on the rate of penetration and the maximum allowable cuttings concentration in the annular space. The following equation was proposed by Guo and Ghalambor (2002) for the required transport velocity:

$$v_{tr} = \frac{\pi d_b^2}{4C_pA}\left(\frac{ROP}{3{,}600}\right) \tag{2.98}$$

where

d_b = bit diameter, in or m
C_p = cuttings concentration, volume fraction
A = annulus cross-sectional area at the depth of interest, in^2 or m^2

For directional well drilling, the minimum required mud velocity for drilling the inclined hole sections is usually considered to be 1.8 times the minimum required mud velocity for drilling the vertical holes. For horizontal well drilling, the minimum required mud velocity for drilling the horizontal hole sections is usually considered to be 1.5 times the minimum required mud velocity for drilling the vertical holes.

Finally, the minimum required mud flow rate in the extreme wellbore geometry can be calculated using

$$q_{min} = 3.1167 v_{min} A \tag{2.99}$$

where

q_{min} = minimum required mud rate, gpm or m^3/min

The constant 3.1167 in U.S. units is 60 in SI units.

SUMMARY

This chapter presented fundamentals for mud hydraulics. Drilling muds are characterized on the basis of their rheological properties. Pressure loss in a conduit depends on fluid properties, flow regime, conduit geometry, and flow rate. The cuttings-carrying capacity of drilling mud is controlled by cutting size and fluid properties.

REFERENCES

Blasius, H., 1913. Das Aehnlichkeitsgesetz bei Reibungsvorgangen in Flussigkeiten. VDL Forsch, 131–137.

Bourgoyne Jr., A.T., Millheim, K.K., Chenevert, M.E., Young Jr., F.S., 1986. Applied Drilling Engineering. SPE Textbook Series, Dallas.

Burkhardt, J.A., 1961. Wellbore pressure surges produced by pipe movement. Trans. AIME 222, 595–605.

Chen, N.H., 1979. An explicit equation for friction factor in pipe. Ind. Eng. Chem. Fund. 18, 296.

Chien, S.F., 1971. Annular velocity for rotary drilling operations. Proceedings of the 5th SPE Conference on Drilling and Rock Mechanics, January 15–16, Austin, pp. 5–16.

Colebrook, C.F., 1938. Turbulent flow in pipes, with particular reference to the transition region between the smooth and rough pipe laws. JICE 11, 133–139.

Dodge, D.G., Metzner, A.B., 1959. Turbulent flow of Non-Newtonian systems. AIChE J 5, 189.

Fang, Q., Guo, B., Ghalambor, A., 2008. Formation of underwater cuttings piles in offshore drilling. SPE Drill. Completion J. (March 2008), 23–28.

Gregory, G.A., Fogarasi, M., 1985. Alternate to standard friction factor equation. Oil Gas J. (April), 120–127.

Guo, B., Ghalambor, A., 2002. Gas Volume Requirements for Underbalanced Drilling Deviated Holes. PennWell Books.

Herschel, W.H., Bulkley, R., 1926. Konsistenzmessungen von Gummi-Benzollosungen. Kolloid-Z 39, 291–300.

Moore, P.L., 1986. Drilling Practices Manual, second ed. PennWell Books.

Walker, R.E., Mayes, T.M., 1975. Design of mud for carrying capacity. JPT (July), Trans. AIME 259, 893–900.

PROBLEMS

2.1 A 12-ppg Bingham plastic fluid with a plastic viscosity of 20 cp and a yield point of 10 lb/100 ft^2 is circulating at 300 gpm in a 7⅞-in-diameter borehole. Determine the flow regime inside a 4½-in OD, 16.60-lb/ft drill pipe (3.826-in ID), and in the drill pipe/hole annulus.

2.2 An 11-ppg Power Law fluid with a flow consistency of 50 eq. cp and flow behavior index of 0.8 is circulating at 300 gpm in an 8½-in-diameter borehole. Determine the flow regime inside a 4½-in OD, 16.60-lb/ft drill pipe (3.826-in ID), and in the drill pipe/hole annulus.

2.3 Predict the parasitic pressure loss under the following conditions:
Total depth: 9,950 ft (3,036 m)
Casing: 9⅝ in, 43.5 lb/ft (8.755-in ID), set at 6,500 ft (1,982 m)
Open hole: 7⅞ in from 6,500 ft to 9,950 ft
Drill pipe: 9,500 ft of 4½ in, 16.6 lb/ft (3.826-in ID)
Drill collar: 450 ft of 6¾-in OD and 2¼-in ID
Surface equipment: Combination 4
Mud weight: 11.5 ppg
Plastic viscosity: 45 cp
Yield point: 10 lb/100 ft^2
Mud flow rate: 350 gpm

2.4 Predict the parasitic pressure loss under the following conditions:
Total depth: 9,950 ft (3,036 m)
Casing: 9⅝ in, 43.5 lb/ft (8.755-in ID), set at 6,500 ft (1,982 m)
Open hole: 8½ in from 6,500 ft to 9,950 ft
Drill pipe: 9,500 ft of 4½ in, 16.6 lb/ft (3.826-in ID)
Drill collar: 450 ft of 6¾-in OD and 2¼-in ID

Surface equipment: Combination 2
Mud weight: 9.5 ppg
Consistency index: 25 eq. cp
Behavior index: 0.85
Mud flow rate: 300 gpm

2.5 Predict the surge pressure generated by a 5½-in casing string under the following conditions, and predict whether the total borehole pressure will exceed the formation fracture gradient. Assume that the casing is effectively "closed" with a float shoe and laminar flow in the annulus.
Casing depth: 4,100 ft (1,250 m)
Fracture gradient: 0.80 psi/ft
Hole diameter: 7⅞ in
Mud weight: 15 ppg
Plastic viscosity: 30 cp
Yield point: 12 lb/100 ft^2
Pipe velocity: –100 ft/min (the negative sign denotes downward velocity)

2.6 Run the computer program *Cuttings Slip Velocity.xls* to perform sensitivity analyses with various cuttings and fluid properties. What can you conclude?

CHAPTER THREE

Mud Pumps

Contents

3.1 INTRODUCTION

Mud pumps are the most important equipment for providing bit hydraulics required for achieving hole cleaning and a high rate of penetration. They should be selected on the basis of flow rates and circulating pressures required at different stages of hole making. Pump power should also be checked. This chapter provides drilling engineers with guidelines for mud pump selection.

3.2 MUD FLOW RATE REQUIREMENTS

The selected mud pump should be capable of providing mud flow rates that are high enough to transport drill cuttings to the surface at all stages of drilling. Since the efficiency of cuttings transport depends on mud properties and mud flow velocity and these parameters change with hole depth, extreme mud properties and extreme annular geometry should be considered.

3.2.1 Extreme Mud Properties

Mud properties that influence the type of pump include mud weight (density) and rheological properties. For Newtonian fluids, viscosity is the only parameter describing fluid rheological characteristics. Plastic viscosity and yield point are the two parameters used to describe the rheological characteristics of Bingham plastic fluids. The consistency and the flow behavior indexes are the two parameters that are utilized to characterize Power Law fluids, also called pseudoplastic fluids in other industries. The consistency index, the flow behavior index, and yield strength are the three parameters that are employed to characterize Herschel-Bulkley fluids. All of these fluid properties can be measured using state-of-the-art instruments used in the oil and gas industry.

In hole cleaning, the properties of the mud affect the settling velocity of drill cuttings in the annulus. To ensure that drilling operations are done safely, the expected ranges of mud properties should be found from mud programs and listed against the hole depths with different borehole geometries. The extreme values in the ranges of properties will be used for estimating the cuttings settling velocity and thus the minimum required mud flow rate from the mud pump.

3.2.2 Extreme Annular Geometry

The minimum required mud flow rate from the mud pump is equal to the minimum required mud velocity times the maximum possible cross-sectional area of annular space during drilling. Therefore, the information of borehole geometry should be known for selecting mud pumps to drill the wells. Figure 3.1 shows a typical borehole geometry diagram. Drill pipe sizes and the extreme mud properties should be marked in the diagram at each level of open hole sizes.

3.2.3 The Minimum Required Flow Rate

The minimum required mud flow rate demanded by the borehole geometry from the mud pump is estimated based on the minimum required mud velocity, which should be higher than the drill cuttings slip velocity. The criterion for the minimum required mud velocity was described in Chapter 2. For mud pump selection, we consider the minimum mud flow rate required for drilling the hole sections of extreme geometries.

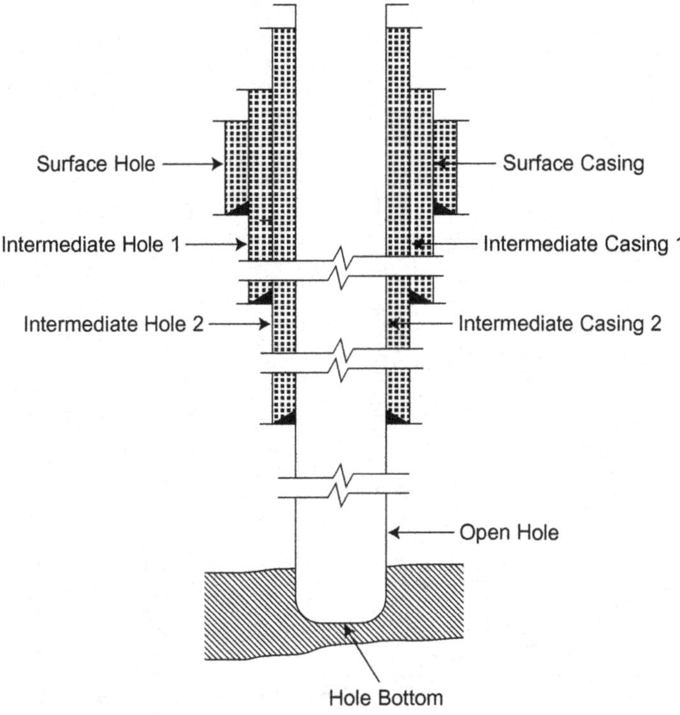

Figure 3.1 A typical borehole geometry diagram.

Illustrative Example 3.1

For the borehole geometry and extreme mud properties given in Figure 3.2, determine the minimum required mud flow rate from the mud pump. Assume that the parameter values in Table 3.1 are realistic.

Solution

The solution was obtained using the computer spreadsheet *Minimum Flow Rates.xls* that is attached to this book. To calculate the minimum flow rare using Table 3.2, (1) select a unit system, (2) update the data in the Input Data column, and (3) click on the Solution button and obtain the result. The result is summarized in Table 3.3. The last column of the table indicates that a mud pump should be selected to be able to provide a minimum mud flow rate of 990 gpm (3.75 m^3/min).

(*Continued*)

Illustrative Example 3.1 *(Continued)*

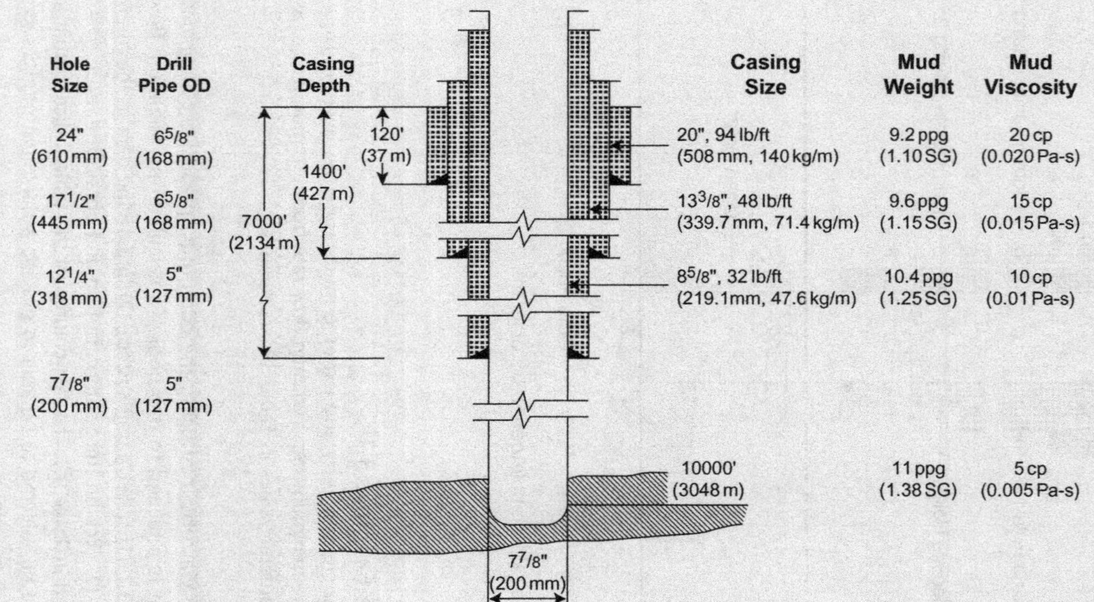

Figure 3.2 Example of borehole geometry with extreme mud properties.

Table 3.1 Rock Properties and Drilling Parameters at Different Hole Depths

Depth		Cuttings Density		Cuttings Sphericity	Rate of Penetration		Rotary Speed	Cuttings Concentration
ft	m	lb/ft^3	g/cc	ball = 1	ft/hr	m/hr	rpm	%
120	37	162	2.60	0.85	90	27.44	70	15
140	427	165	2.65	0.8	70	21.34	60	10
7,000	2,134	168	2.70	0.75	50	15.24	50	8
10,000	3,048	172	2.75	0.7	60	18.29	40	5

Table 3.2 Computer Spreadsheet *Minimum Flow Rates.xls*

Input Data	**U.S. Units**	**SI Units**
Cuttings specific gravity	2.6	water = 1
Particle sphericity	0.85	ball = 1
Drilling fluid viscosity	20 cp	Pa-s
Drilling fluid density	9.2 ppg	g/cc
Annulus OD	24 in	mm
Annulus ID	6.625 in	mm
Rate of penetration	90 ft/hr	m/hr
Rotary speed	70 rpm	rpm
Cuttings concentration	15%	%
Solution		
Cuttings equivalent diameter	0.26 in	
$A' = 2.2954 - 2.2626\psi + 4.4395\psi^2 - 2.9825\psi^3$	1.7481	
$B' = -0.4193 - 1.9014\psi + 3.3416\psi^2 - 2.0409\psi^3$	-0.8746	
$C' = 0.1117 + 0.0553\psi - 0.1468\psi^2 + 0.1145\psi^3$	0.1230	
$N_{\mathrm{Re}P} = \dfrac{928\rho_f v_{sl} d_s}{\mu}$	64	
$f_p = 10\wedge\left(A' + B'\log(N_{\mathrm{Re}P}) + C'[\log(N_{\mathrm{Re}P})]^2\right)$	3.7215	
$v_{sl} = 1.89\sqrt{\dfrac{d_s}{f_p}\left(\dfrac{\rho_s - 7.48\rho_f}{7.48\rho_f}\right)}$	0.58 ft/s	
$A = \dfrac{\pi(d_a^2 - d_b^2)}{4}$	418 in^2	
$v_{tr} = \dfrac{\pi d_b^2}{4C_pA}\left(\dfrac{ROP}{3{,}600}\right)$	0.18 ft/s	
$v_{min} = v_{sl} + v_{tr}$	0.76 ft/s	
$q_{min} = 3.1167 v_{min} A$	**990** gpm	

(Continued)

Illustrative Example 3.1 *(Continued)*

Table 3.3 Summary of Calculated Results

Depth		Cuttings Size		Slip Velocity		Transport Velocity		Mud Velocity		Mud Flow Rate	
ft	m	in	mm	ft/s	m/s	ft/s	m/s	ft/s	m/s	gpm	m^3/min
120	37	0.26	6.53	0.58	0.18	0.18	0.055	0.76	0.23	990	3.75
140	427	0.23	5.93	0.49	0.15	0.23	0.069	0.72	0.22	460	1.74
7,000	2,134	0.20	5.08	0.41	0.13	0.21	0.064	0.62	0.19	189	0.72
10,000	3,048	0.30	7.62	0.55	0.17	0.56	0.170	1.11	0.34	100	0.38

3.3 PRESSURE REQUIREMENTS

The selected mud pump should also be capable of providing pressure that is strong enough to overcome the total pressure loss and pressure drop at the bit in the circulating system at the total hole depth. The pressure loss depends on the mud properties, the drill string configuration, the borehole geometry, and the mud flow rate. The pressure drop at the bit should be optimized based on the total pressure loss in the system to maximize bit hydraulics. Therefore, extreme borehole architecture and condition should be considered.

3.3.1 Extreme Borehole Configurations

Maximum pressure loss normally occurs when the total hole depth is reached. At this point, the drill string and the open hole section assume their longest values. The borehole configuration is shown in Figure 3.3. To perform pressure loss calculations, it is convenient to put the dimension (lengths and diameters) data along the circulating path in the graph.

3.3.2 Extreme Borehole Conditions

The maximum circulating pressure normally occurs at the total depth with extreme borehole conditions. These conditions include the use of a mud flow rate higher than normal to clean the hole. Different mud properties are used, and the mud weight is increased before tripping out the drill string. These extreme parameter values should be marked in the borehole configuration graph for pressure loss calculations.

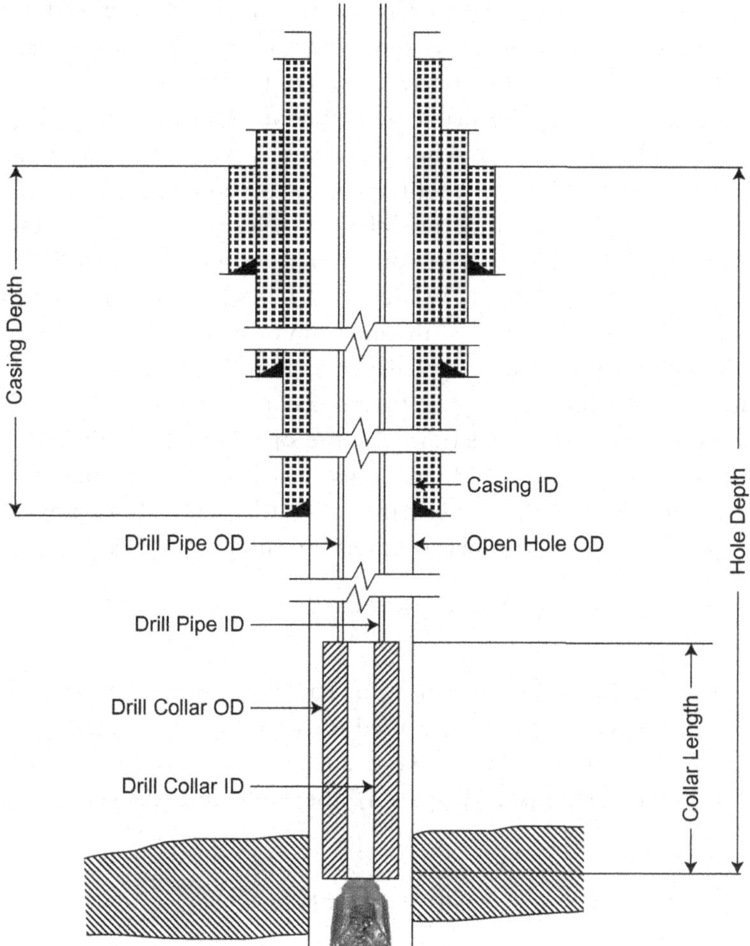

Figure 3.3 Borehole configuration at the total depth.

3.3.3 Circulating Pressure

The maximum expected circulating pressure is the total frictional pressure loss and pressure drop at the bit at the total hole depth. The frictional pressure loss depends on the fluid properties, the flow velocity, the flow regime, and the length of the flow path. Under normal drilling conditions, turbulent flow exists inside the surface equipment, the drill pipe, the drill collar, and the annulus outside the drill collar. Laminar flow normally exists in the annulus outside the drill pipe. The pressure loss in turbulent flow is usually higher than that in laminar flow. For the purpose of pump selection, assuming turbulent flow throughout the circulating system will result in conservative values of pressure losses.

This section presents the analytical method used for predicting the pressure losses in the drill string and in the annulus, as well as considerations for pressure drop at the bit. The length of the surface equipment is considered to be a small fraction of that of the drill string. Necessary hydraulics models were presented in Chapter 2. For directional and horizontal drilling, the pressure losses through the MWD and LWD tools are considered to be negligible. The pressure drop at the mud motor is considered as a specific value between 200 psi and 600 psi, depending on motor size.

However, the pressure drop at the bit is not calculated with Eq. (2.64) in the process of pump selection. For the optimum bit hydraulics, the pressure drop at the bit should be selected based on the total pressure loss in the system. According to the maximum bit hydraulic horsepower criterion (see Chapter 4), the following relation should be held:

$$\Delta p_b = \frac{m}{m+1} p_p \tag{3.1}$$

where p_p is the pump pressure in psi or Pa, and m is the flow rate exponent. If the Blasius correlation is used for friction factor determination, Eq. (3.2) shows $m = 1.75$. However, according to the maximum jet impact force criterion (see Chapter 4), the following relation should be held:

$$\Delta p_b = \frac{m}{m+2} p_p \tag{3.2}$$

3.3.4 The Minimum Required Pressure

The minimum required pump pressure is expressed as

$$p_p = \Delta p_d + \Delta p_b \tag{3.3}$$

where

Δp_d = the total frictional pressure loss (parasitic pressure) in psi or N/m^2—that is,

$$\Delta p_d = \sum_{i=1}^{n} p_{fi}$$

Combining Eqs. (3.1) and (3.3), the following relation is derived for the maximum bit hydraulic horsepower criterion:

$$\Delta p_d = \frac{1}{m+1} p_p \tag{3.4}$$

Combining Eqs. (3.2) and (3.3), the following relation is derived for the maximum jet impact force criterion:

$$\Delta p_d = \frac{2}{m+2} p_p \tag{3.5}$$

From which the expressions for the required pump pressure are

$$p_p = (m+1)\Delta p_d \tag{3.6}$$

and

$$p_p = \frac{m+2}{2}\Delta p_d \tag{3.7}$$

for the maximum bit hydraulic horsepower criterion and the maximum jet impact force criterion, respectively.

Illustrative Example 3.2

For the data in Illustrative Example 3.1 and the additional data given in Figure 3.4, determine the minimum required pump pressure. Assume the maximum mud weight of 12 ppg (1,440 kg/m^3), the maximum plastic viscosity of 15 cp (0.015 Pa-s), the maximum yield point of 10 lb/100 ft^2 (4.78 Pa), pipe wall roughness of 0.00025 in. (0.00635 mm), pressure drop at the mud motor of 200 psi (1,379 kPa), and a mud flow rate of 300 gpm (1.14 m^3/min).

Solution

This problem is solved using the spreadsheet program *Pump Pressure.xls* that is attached to this book. To calculate the required pump pressure using Table 3.4, (1) select a unit system, (2) update the data in the Input Data column, and (3) click on the Solution button and obtain the result. The result is summarized in Table 3.5 for the maximum bit hydraulic horsepower criterion. It indicates that the required pressure is 3,461 psi (23.86 MPa).

(*Continued*)

Illustrative Example 3.2 *(Continued)*

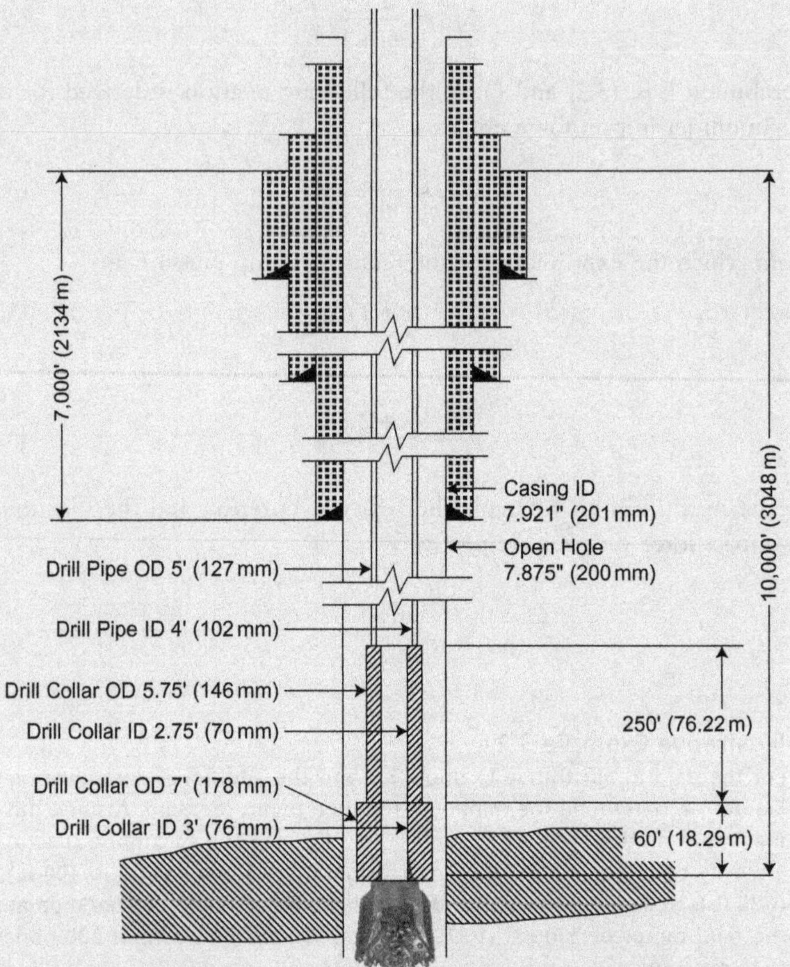

Figure 3.4 Example of borehole configuration at the total depth.

Table 3.4 Part of the spreadsheet program *Pump Pressure.xls*

Input Data	1
Hole depth	10,000 ft
Open hole diameter	7.875 in
Open hole roughness	0.05 in
Cased hole depth	7,000 ft
Cased hole diameter	7.921 in

Table 3.4 (*Continued*)

Input Data	1	
Pipe roughness	0.0025 in	
Length of drill collar 1	60 ft	
OD of collar 1	7.000 in	
ID of collar 1	3.000 in	
Length of drill collar 2	0 ft	
OD of collar 2	6.250 in	
ID of collar 2	2.500 in	
Length of drill collar 3	250 ft	
OD of collar 3	5.750 in	
ID of collar 3	3.000 in	
Drill pipe OD	5.000 in	
Drill pipe ID	4.000 in	
Mud weight	11 ppg	
Plastic viscosity	5 cp	
Yield point	5 lb/100 ft^2	
Mud flow rate	300 gpm	
Pressure drop at mud motor	200 psi	
Flow rate exponent	1.75 m	
Solution	**Casing Pipe**	**Hole Pipe**
$\overline{v} = \frac{q}{2.448(d_o^2 - d_i^2)}$	3.25 ft/s	3.31
$d_e = 0.816(d_o - d_i)$	2.38 in	2.35
$\mu_a = \mu p + \frac{6.66\tau_y d_e}{\overline{v}}$	29.44 cp	28.60
$N_{\text{Re}} = \frac{928\rho_f \overline{v} d_e}{\mu_a}$	2,683	2,719
$f = \left(-4\log\left\{\frac{\varepsilon}{3.7065} - \frac{5.0452}{N_{\text{Re}}}\log\left[\frac{\varepsilon^{1.1098}}{2.8257} + \left(\frac{7.149}{N_{\text{Re}}}\right)^{0.8981}\right]\right\}\right)^{-2}$	0.0137	0.0137
$\frac{dp_f}{dL} = \frac{f\rho_f \overline{v}^2}{25.8d}\arcsin\theta$	0.026 psi/ft	0.027
$p_f = \left(\frac{dp_f}{dL}\right)L$	181 psi	73
$\Delta p_d = \sum_{i=1}^{n} p_{fi}$	1,259 psi	
$p_p = (m+1)\Big)\Delta p_d$	**3,461** psi	
$p_p = \frac{m+2}{2}\Delta p_d \pi$	**2,360** psi	

(*Continued*)

Illustrative Example 3.2 *(Continued)*

Table 3.5 Summary of Calculated Pressures

Equipment	Pressure Loss/Drop psi	MPa
Inside drill pipe	636	4.39
Inside top drill collar	96	0.66
Inside mid-drill collar	0	0
Inside bottom drill collar	15	0.10
Motor	200	1.38
Bit nozzles	2,202	15.19
Outside bottom drill collar	42	0.29
Outside mid-drill collar	0	0
Outside top drill collar	14	0.10
Drill pipe open hole annulus	73	0.50
Drill pipe cased hole annulus	181	1.25
Total circulation pressure	3,461	23.86

3.4 HORSEPOWER REQUIREMENTS

In rotary drilling, the engines that supply power are rated on output horsepower, sometimes called brake horsepower. Fluid pumps that receive power are rated on the basis of input horsepower. For this reason, a 1,600-hp pump classification means that the horsepower fed into the pump should not exceed 1,600. Output horsepower from pumps used in rotary drilling is determined from charts of maximum permissible surface pressure and maximum circulation rate.

Mud pumps are rated by horsepower P_H and the maximum working pressure p_{pm}. Figure 3.5 shows a theoretical pump performance curve. The mud hydraulic horsepower from the pump is expressed as (Moore, 1986)

$$P_h = \frac{qp}{1,714} \tag{3.8}$$

where

P_h = hydraulic horsepower, hp
q = mud flow rate, gpm or m^3/min
p = pump pressure, psi or kPa

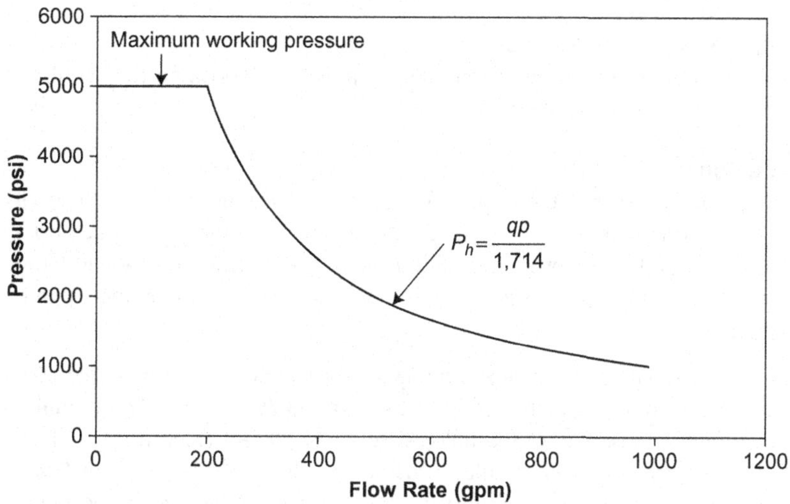

Figure 3.5 A theoretical pump performance curve.

The constant 1,714 in U.S. units is 44.14 in SI units.

For a given pump having a horsepower rating P_H, the value of the right-hand side of Eq. (3.8) should not exceed P_H; that is, $P_h < P_H$. If a pump runs at the maximum working pressure p_{pm}, the maximum available flow rate is expressed as

$$q_{max} = \frac{1{,}714 E_p P_H}{p_{pm}} \tag{3.9}$$

where

q_{max} = maximum mud flow rate, gpm or m^3/min
P_H = Horsepower rating of pump, hp
E_p = pump efficiency, dimensionless
p_{pm} = maximum working pressure of pump, psi or MPa

If a pump runs at a flow rate $q < q_{max}$, the maximum available pump pressure is expressed as

$$p_{max} = \frac{1{,}714 E_p P_H}{q} \tag{3.10}$$

However, the pump pressure should always be kept lower than the maximum working pressure—that is, $p_{max} < p_{pm}$.

Illustrative Example 3.3

For the data in Illustrative Examples 3.1 and 3.2, determine the required horsepower rating of the pump.

Solution

The pump should be able to provide adequate horsepower while drilling all hole sections. The extreme hole conditions occur when the surface hole and the total hole depth are drilled. Drilling the surface hole requires the highest mud flow, and drilling at the total depth requires the highest pump pressure.

Surface Hole Drilling. Illustrative Example 3.1 shows that the minimum required flow rate to drill the surface hole is 990 gpm (3.75 m^3/min). The required pressure at the bottom of the hole section with 60 feet (18.3 m) of a 7-inch (178 mm) drill collar is calculated using the spreadsheet program *Pump Pressure.xls*. The result of the pressure loss is 364 psi (2,509 kPa). Considering a pressure drop at the bit of twice the pressure loss, the circulating pressure will be 1,092 psi (7,529 kPa). Substituting these data into Eq. (3.9) gives

$$P_h = \frac{(990)(1{,}092)}{1{,}714} = 631\,\text{hp}$$

Drilling at the Total Depth. Using the flow rate of 350 gpm (1.325 m^3/min) and the required pressure of 3,461 psi (23,863 kPa) calculated with the spreadsheet program *Pump Pressure.xls*, Eq. (3.9) gives

$$P_h = \frac{(350)(3{,}641)}{1{,}714} = 743\,\text{hp}$$

Therefore, the minimum required horsepower rating of the pump is 743 hp.

3.5 CAPACITIES OF MUD PUMPS

The two types of piston strokes in mud pumps are the single-action piston stroke and the double-action piston stroke, which are shown in Figures 3.6 and 3.7. The double-action stroke is used for duplex (two pistons) pumps. The single-action stroke is used for triplex pumps. Normally, duplex pumps can handle higher flow rates, and triplex pumps can provide higher pressure. The discharged flow rate depends on several parameters, including the liner size, the rod size, the stroke length, the

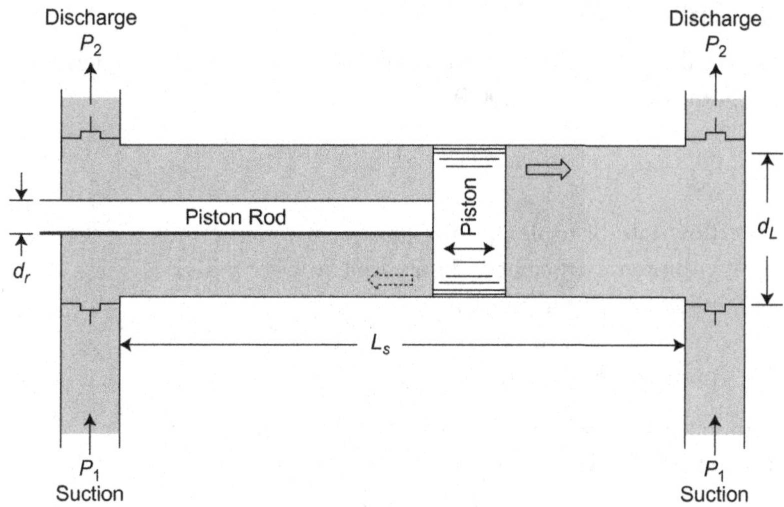

Figure 3.6 Double-action stroke in a duplex pump.

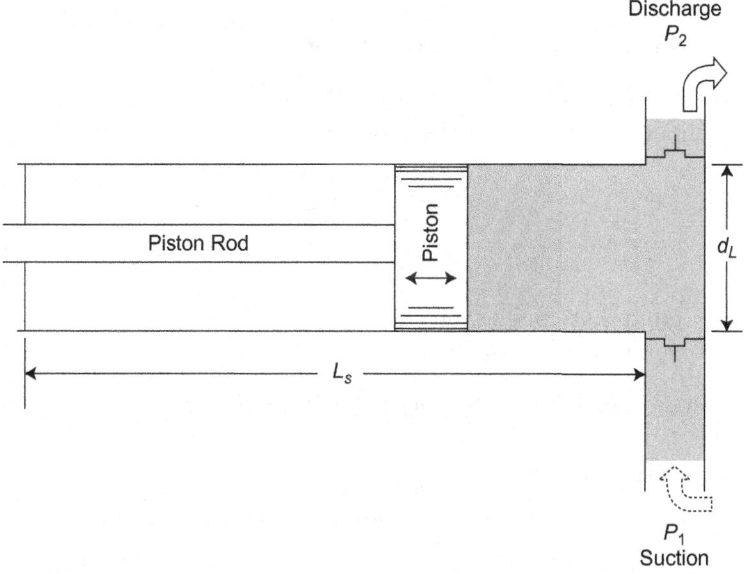

Figure 3.7 Single-action stroke in a triplex pump.

pumping speed, and the volumetric efficiency. The rod size changes with the size of the liner. The pumping speed can be adjusted if diesel engines or DC motors are used as the prime movers. The volumetric efficiency varies with the fluid properties.

3.5.1 Triplex Pumps

Geometrical analysis allows for the following equation to be derived for triplex pumps (Guo et al., 2007):

$$q_T = 0.01 e_v d^2 l N \tag{3.11}$$

where

q_T = flow rate of triplex pump, gpm or m^3/min
e_v = volumetric efficiency, dimensionless
d = piston diameter, in or m
l = stroke length, in or m
N = pumping speed, spm

The constant 0.01 in U.S. units is 2.3066 in SI units.

The pumped volume per stroke is

$$q_S = \frac{e_v d^2 l}{4{,}118} \tag{3.12}$$

where

q_S = pumped volume per stroke, bbl or m^3

The constant 4,118 in U.S. units is 0.4201 in SI units.

The input horsepower needed from the prime mover is expressed as

$$HP = \frac{p d^2 l N}{168{,}067 e_m} \tag{3.13}$$

where

HP = pump horsepower, hp
e_m = pump mechanical efficiency, dimensionless

The constant 168,067 in U.S. units is 18.98 in SI units.

3.5.2 Duplex Pumps

Geometrical analysis allows for the following equation to be derived for duplex pumps (Guo et al., 2007):

$$q_D = 0.0068 e_v (2d_1^2 - d_2^2) l N \tag{3.14}$$

where

q_D = flow rate of duplex pump, gpm or m^3/min
d_1 = piston diameter, in or mm
d_2 = rod diameter, in or mm

The constant 0.0068 in U.S. units is 1.57×10^{-9} in SI units.

The pumped volume per stroke is

$$q_S = \frac{e_v d^2 l}{5{,}912} \tag{3.15}$$

where

q_S = pumped volume per stroke, bbl or m^3

The constant 5,912 in U.S. units is 0.6069 in SI units.

The input horsepower needed from the prime mover is expressed as

$$HP = \frac{p(2d_1^2 - d_2^2)lN}{252{,}101 e_m} \tag{3.16}$$

The constant 252,101 in U.S. units is 2.8×10^{10} in SI units.

Illustrative Example 3.4

Two identical pumps are considered for drilling the well shown in Illustrative Example 3.2. These pumps are the TSC WF700 triplex pumps, each having a stroke length of 8.5 in. (0.216 m). Table 3.6 provides the pump specification data given by the manufacturer.

The pump can run at the maximum speed of 150 spm with piston diameters from 4 in. to 7 in. (0.108 m to 0.178 m). Assuming that the volumetric efficiency and mechanical efficiency are 0.95 and 0.9, respectively, if the flow rate of 350 gpm (1.325 m^3/min) is desired with 4.5-in. (0.114 m) liners, what is the required pump speed? What is the total input horsepower needed from the prime movers?

Solution

Equation (3.11) gives

$$N = \frac{q_T}{0.01 e_v d^2 l}$$
$$= \frac{350/2}{0.01(0.95)(4.5)^2(8.5)}$$
$$= 107 \text{ spm}$$

(Continued)

Illustrative Example 3.4 *(Continued)*

Table 3.6 Specifications of TSC WF700 Triplex Pumps

		Pinion RPM	30	60	160	260	360	460	560	660	760
Performance Characteristics		Pinion lb/ft	4912								
		Pinion HP	28	56	150	243	337	430	524	617	711
Piston Diameter (in)	Pressure (psi)	Speed (spm)	6	12	32	51	71	91	111	131	150
4	5,253	Flow rate (gpm)	8.2	16.5	43.9	71.4	98.9	126	154	181	209
4.5	4,151		10.4	20.9	55.6	90.4	125	160	195	229	264
5	3,362		12.9	25.7	68.7	112	155	197	240	283	326
5.5	2,779		15.6	31.2	83.1	135	187	239	291	343	395
6	2,335		18.5	37.1	98.9	161	223	284	346	408	470
6.5	1,989		21.8	43.5	116	189	261	334	406	479	551
7	1,715		25.2	50.5	135	219	303	387	471	555	639

Equation (3.13) gives horsepower for each pump:

$$HP = \frac{pd^2 lN}{168{,}067e_m}$$

$$= \frac{(3461)(4.5)^2(8.5)(107)}{168{,}067(0.90)}$$

$$= 421 \text{ hp}$$

Therefore, the total input horsepower needed from the prime movers is (2)(421) = 842 hp. The safety factor in horsepower is (700)/(421) = 1.67. According to Eq. (3.16), the maximum allowable pump pressure is calculated to be 5,862 psi (40 MPa). However, the manufacturer-recommended maximum working pressure for the 4.5-in. liner is 4,151 psi. The safety factor in pressure is (4,151)/(3,461) = 1.2.

SUMMARY

This chapter presented theory and procedures for selecting mud pumps. Extreme borehole geometries and mud properties should be considered for calculating the minimum required flow rate, pressure, and horsepower. Safety factors should be applied.

REFERENCES

Guo, B., Lyons, W.C., Ghalambor, A., 2007. Petroleum Production Engineering, A Computer-Assisted Approach. Elsevier.

Moore, P.L., 1986. Drilling Practices Manual, second ed. PennWell Books.

PROBLEMS

3.1 Solve the problem in Illustrative Example 3.1 with a cuttings concentration value of 10% in all hole sections.

3.2 Solve the problem in Illustrative Example 3.2 with a mud flow rate of 300 gpm.

3.3 Solve the problem in Illustrative Example 3.3 with a mud flow rate of 300 gpm.

3.4 Solve the problem in Illustrative Example 3.4 with a liner size of 4 in.

CHAPTER FOUR

Mud Hydraulics Optimization

Contents

4.1 INTRODUCTION

Drilling hydraulics is considered the most important factor in drilling performance. The rate of penetration can be significantly increased using state-of-the-art techniques for hydraulics optimization to minimize drilling cost. The goal of the optimization is to make the maximum usage of the pump's power to help the bit to drill at maximum efficiency. This is achieved by minimizing the energy loss due to friction in the circulating system and use the saved energy to improve bit hydraulics. Starting with fluid rheology basics, this chapter provides drilling engineers with a practical approach to optimizing drilling hydraulics.

4.2 THE CRITERIA OF HYDRAULICS OPTIMIZATION

An operator's primary concern with drilling fluid hydraulics is achieving adequate hole cleaning below the bit. This is important for the following reasons:

- The rate of penetration (ROP) depends on hole cleaning below the bit.
- Some bits can be damaged due to overheating if cuttings accumulate below them.

- Poor cleaning below the bit hinders the detection of changes in formation properties that otherwise can be identified from the rate of penetration.

Achieving an adequate level of hole cleaning requires maximum use of the power from the mud pumps on the bit hydraulics. This means the maximal use of pump pressure and flow rate.

For given pumps with fixed horsepower, the available pump pressure and flow rate are determined by the size of the liner (piston) used. Therefore, the optimum selection of pressure and flow rate is not straightforward. This section describes the criteria and procedure used in the drilling industry for optimizing mud hydraulics to help achieve the maximum rate of penetration.

There are different theories regarding the mechanism of hole cleaning. Different design criteria have been used to optimize fluid hydraulics for maximizing hole cleaning and thus the rate of penetration. These criteria include the maximum bit hydraulic horsepower, the maximum bit hydraulic impact force, and the maximum bit jet velocity.

4.2.1 The Maximum Bit Hydraulic Horsepower Criterion

Horsepower is defined as the rate of doing work. One horsepower is equivalent to 33,000 foot-pounds of work done in one minute. This definition is universal, and, other than changes in units, it applies all over the world. The maximum bit hydraulic horsepower criterion may be stated as follows: Within the maximum available pump pressure, mud flow rate and nozzle size should be chosen so the bit will gain the maximum possible horsepower to clean the bottom hole.

Speer (1958) pointed out that the effectiveness of jet bits could be improved by increasing the hydraulic power of the pump. He reasoned that the penetration rate would increase with hydraulic horsepower until the cuttings were removed as fast as they were generated. After this "perfect cleaning" level was achieved, there should be no further increase in the penetration rate with hydraulic power. Shortly after Speer published his findings, several authors pointed out that due to the frictional pressure loss in the drill string and annulus, the hydraulic power developed at the bottom of the hole is different from the hydraulic power developed by the pump. They concluded that bit horsepower rather than pump horsepower is the important parameter. Furthermore, it was concluded that bit horsepower is not necessarily maximized by operating the pump at the maximum possible horsepower. The conditions for maximum bit horsepower were derived by Kendall and Goins (1960).

The pressure drop at the bit is expressed as

$$\Delta p_b = p_p - cq^m \tag{4.1}$$

Substituting this expression into Eq. (2.88) gives

$$P_{Hb} = \frac{q(p_p - cq^m)}{1{,}714} \tag{4.2}$$

Using calculus to determine the flow rate at which the bit horsepower is a maximum gives

$$\frac{dp_{Hb}}{dq} = \frac{p_p - (m+1)cq^m}{1{,}714} = 0 \tag{4.3}$$

Solving for the root of this equation yields

$$p_p = (m+1)cq^m = (m+1)\Delta p_d \tag{4.4}$$

or

$$\Delta p_d = \frac{p_p}{m+1} \tag{4.5}$$

It can be shown that $\frac{d^2 p_{Hb}}{dq^2} < 0$ at this root, so the root corresponds to a maximum. Thus, bit hydraulic horsepower is a maximum when the parasitic pressure loss is $\frac{1}{m+1}$ times the pump pressure. Since

$$\Delta p_b = p_p - \Delta p_d = p_p - \frac{p_p}{m+1} = \frac{m}{m+1} p_p \tag{4.6}$$

bit hydraulic horsepower is a maximum when the pressure drop at that bit is $\frac{m}{m+1}$ times the pump pressure.

4.2.2 The Maximum Jet Impact Force Criterion

The maximum jet impact force criterion may be stated as follows: Within the maximum available pump pressure, the mud flow rate and the nozzle size should be chosen so the bit will exert the maximum possible jet impact force to clean the bottomhole.

The conditions for maximum jet impact force were also derived by Kendall and Goins (1960). Substituting Eq. (4.1) into Eq. (2.89) gives

$$F_j = 0.01823 C_d q \sqrt{\rho(p_p - cq^m)} \tag{4.7}$$

Using calculus to determine the flow rate at which the bit impact force is a maximum gives

$$\frac{dF_j}{dq} = \frac{0.009115C_d[2\rho\Delta p_p q - (m+2)\rho c q^{m+1}]}{\sqrt{\rho\Delta p_p q^2 - \rho c q^{m+2}}} = 0 \quad (4.8)$$

Solving for the root of this equation yields

$$\Delta p_d = \frac{2p_p}{m+2} \quad (4.9)$$

It can be shown that $\frac{d^2F_j}{dq^2} < 0$ at this root, so the root corresponds to a maximum. Thus, the jet impact force is a maximum when the parasitic pressure loss is $\frac{2}{m+2}$ times the pump pressure. Since

$$\Delta p_b = p_p - \Delta p_d = p_p - \frac{2p_p}{m+2} = \frac{m}{m+2}p_p \quad (4.10)$$

the bit jet impact force is a maximum when the pressure drop at the bit is $\frac{m}{m+2}$ times the pump pressure.

4.2.3 The Maximum Nozzle Velocity Criterion

The maximum nozzle velocity criterion may be stated as follows: Within the maximum available pump pressure, the mud flow rate and the nozzle size should be chosen so the bit will create the maximum possible jet velocity to clean the bottomhole.

Substituting Eq. (4.1) into Eq. (2.87) gives

$$v_n = C_d\sqrt{\frac{p_p - cq^m}{8.074\times 10^{-4}\rho}} \quad (4.11)$$

This equation implies that the nozzle velocity can be increased by reducing the flow rate so the parasitic pressure loss is reduced. In field applications, the flow rate is set to the minimum flow rate determined by the minimum annular velocity required to lift cuttings.

4.2.4 Bit Hydraulics

Regarding the question of which criterion is the best for optimizing bit hydraulics, most people use the maximum bit hydraulic horsepower or the maximum bit hydraulic impact force criterion at shallow to middle depths and then shift to the maximum nozzle velocity at deeper depths.

Between the maximum bit hydraulic horsepower and the maximum bit hydraulic impact force criteria, neither criterion has been proved better in all cases because there is little difference in the application of the two procedures. If the jet impact force is a maximum, the hydraulic horsepower will be within 90% of the maximum and vice versa. Another argument is that in many cases bits provide higher than required hydraulics, so the effects of design using different criteria are masked.

The concept of bit hydraulic horsepower was introduced as a design criterion in the early 1950s. It is a measure of the work required to squeeze mud through the bit nozzles. This work is related to the removal of cuttings from below the bit. Bit hydraulic horsepower is the most common design procedure, probably because it was used first.

The concept of hydraulic impact force as a design criterion was introduced in the mid-1950s. Hydraulic impact force is a measure of the force exerted by the fluid at the exits of the bit nozzles. This fluid force cleans the bottomhole by direct erosion and by cross flow beneath the bit. Hydraulic impact force below the bit seems more logical than the bit hydraulic horsepower when considering design procedures for bottomhole cleaning. Rock bits with jet nozzles extended closer to the bottom of the hole are widely used. Both laboratory and field tests have shown better bottomhole cleaning with extended bit nozzles (Sutko, 1970). Since extending the nozzles does not change the bit hydraulic horsepower but does change the hydraulic impact force on the bottom of the hole, it is believed that the latter relates more directly to hole cleaning.

4.2.5 Economical Bit Hydraulics

Both bit hydraulic horsepower and hydraulic impact force criteria are widely used in designing mud hydraulics programs. The argument about which design criterion to use may be moot because either can be utilized to optimize bottomhole cleaning requirements. Drilling tests in actual drilling operations determine the optimum cleaning requirements. Therefore, if the bottomhole cleaning requirements are determined using bit hydraulic horsepower, then bit hydraulic horsepower should be the design base. The same holds true for hydraulic impact force.

In formations of normal hardness where no specific breaking point is present, the amount of bottomhole cleaning necessary may be determined directly in field operations. It may be difficult to determine the hole cleaning required for maximum penetration rates in very soft formations. In these formations, maximum penetration rates are achieved with maximum bottomhole jetting action. Therefore, the problem is one of using the maximum jetting action that is economically feasible. Economic feasibility depends on the maximum penetration rates possible, the hole conditions, and other factors such as connection time and the maintenance of support equipment.

When high-capacity pumps are available, it is possible to achieve a higher level of bit hydraulics (horsepower, impact force, or nozzle velocity) than is needed to clean the hole bottom adequately. Using higher than needed bit hydraulics not only is wasteful but also can be harmful. This is because the high flow velocity in the system can result in borehole and pipe erosion as well as premature failure of the pump's parts.

It is important to know that pump maintenance costs go up as pump pressures are increased. Showing a direct mathematical relationship between pump pressures and maintenance costs is difficult because so many other variables also have direct impacts on pump maintenance expense. In fact, pump maintenance costs rise much faster than the increase in pump pressures. For instance, the pump maintenance cost is often more than doubled when the pump pressure increases from 2,500 to 3,000 psi (17,006 to 20,407 kPa). Precise numbers for specific rigs or operations must be determined based on field operation conditions.

If the hole cleaning needs can be established from penetration rate data taken in similar lithology under conditions of varying bit hydraulics, the pump energy input should be reduced by decreasing the flow rate until the desired level of bit hydraulics can be obtained if the pump is operated at the maximum allowable pressure. This same logic could be applied using either hydraulic horsepower or impact force as the hydraulic parameter.

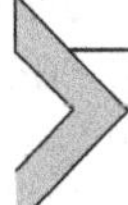

4.3 THE OPTIMUM DESIGN OF A HYDRAULICS PROGRAM

For a drilling operation with given mud pumps and mud programs, a well hydraulics program is defined as a complete procedure of changing the pump liner, the flow rate, and the nozzle size according to the depth of drilling. Drilling performance can always be improved by optimizing the hydraulics program.

4.3.1 Selecting the Liner Size

As shown in Table 3.6, pump manufacturers provide specifications of pumps that indicate the maximum pressures and ranges of flow rates for different liner sizes. The liner size to use to drill a specific hole section specified by the bit size should provide an adequate flow rate for carrying up drill cuttings and sufficient pressure to drill the section completely.

In field drilling operations, there is often confusion about the maximum pump pressure and the maximum circulation rates. Pump manufacturers publish what are called "maximum liner ratings" and "maximum circulation rates" for specific pumps. These maximums, however, are seldom used in drilling operations. The manufacturer's published maximum pressure is based on the maximum permissible force on the power end bearings.

In actual drilling operations, the maximum pump pressure is rarely achieved. Many arbitrary standards are used. One common standard utilizes a fixed percentage of the maximum liner rating pressure. Most operators would not exceed 90% of the liner rating, which is equivalent to a safety factor of 1.11. Some rig operators ignore the safety factor completely and simply specify a maximum surface pump pressure.

Because the combination of flow rate and pressure is limited by pump horsepower, the high flow rate is available from a pump with a scarifying (reduced) available pump pressure. On the other hand, high pressure is available from a pump with a scarifying (reduced) available flow rate. The procedure for selection of liner size is as follows:

1. For the given borehole geometry and fluid properties, calculate the minimum flow rate required to transport drilling cuttings to the surface (see Chapter 2).
2. Design a flow rate based on the minimum required flow rate with a sufficient safety factor.
3. Look up the manufacturer's pump specifications table to select a liner size that meets the designed flow rate from two pumps.
4. Based on the designed flow rate, calculate the expected total parasitic pressure loss at the total depth of the section.
5. Calculate the required pump pressure based on the total parasitic pressure loss and the selected criterion for hydraulics optimization.
6. Look up the manufacturer's pump specifications table to check the maximum allowable pressure against the calculated pump pressure.
7. If the maximum allowable pressure is higher than the calculated pump pressure, this liner size can be selected. Otherwise, go back to step 3 and consider a smaller liner size.

Illustrative Example 4.1

For the data in Illustrative Example 3.2, select a liner size for two TSC WF700 Triplex pumps. Assume the flow rate exponent $m = 1.75$ and the maximum bit hydraulic horsepower criterion. Additional data are given as follows:

Cuttings specific gravity: 2.7 (water = 1)
Particle sphericity: 0.85 (ball = 1)
Rate of penetration: 90 ft/hr
Rotary speed: 70 rpm
Cuttings concentration: 10%

Solution

1. For the given borehole geometry and fluid properties, the minimum flow rate required to transport drill cuttings to the surface is calculated using computer program *Minimum Flow Rates.xls*. The result is

$$q_{min} = 137 \text{ gpm} \left(0.52\text{m}^3/\text{min}\right)$$

2. The designed flow rate in Illustrative Example 3.2 is

$$q_{des} = 300 \text{ gpm} \left(1.14\text{m}^3/\text{min}\right)$$

   The safety factor is 2.2.

3. The manufacturer's pump specifications in Table 3.6 show that several liner sizes meet this designed flow rate. The liner size of 5 in (0.127 m) can be a good candidate. Two pumps with this liner size will provide mud flow rates between (2)(155) = 310 gpm (1.18 m^3/min) and (2)(326) = 652 gpm (2.48 m^3/min) at pumping speeds from 71 spm to 150 spm.
4. Based on the designed flow rate of 300 gpm (1.14 m^3/min), the expected total parasitic pressure loss at a depth of 9,950 ft (3,036 m) was calculated in Illustrative Example 3.2 as

$$\Delta p_d = 1{,}087 \text{ psi} \left(7{,}394 \text{ kPa}\right)$$

5. Equation (3.6) gives

$$p_p = (m+1)\Delta p_d = (1.75+1)(1{,}087) = 2{,}989 \text{ psi} \left(20{,}333 \text{ kPa}\right)$$

6. The manufacturer's pump specifications in Table 3.6 show that the maximum allowable pressure for the 5-in (0.127-m) liner is 3,362 psi (22,869 kPa). This pressure is greater than the calculated pump pressure of 2,989 psi (20,333 kPa).
7. The pressure safety factor is (3,362)/(2,989) = 1.125. Thus, the liner size of 5 in (0.127 m) can be selected for drilling this hole section.

4.3.2 Selecting the Flow Rate and the Bit Nozzle Size

For a given hole section to be drilled with a selected pump liner size, the optimum mud flow rates and the nozzle sizes should be designed for drilling at various depths until the end of the section. Traditionally, the flow rate design has been performed graphically. It is a routine practice today to carry out the design with computer programs.

The following procedure for selecting the optimum mud flow rate and nozzle size is valid for the maximum bit hydraulic power and the maximum jet impact force criteria. Based on the information presented earlier in this chapter, it is clear that the optimum bit-to-pump pressure ratio is $\frac{\Delta p_b}{p_p} = \frac{m}{m+1}$ for the maximum bit hydraulic power criterion and $\frac{\Delta p_b}{p_p} = \frac{m}{m+2}$ for the maximum jet impact force criterion. From a practical standpoint, it is not always desirable to maintain the optimum pressure ratio. At shallow depths, the flow rate usually is held constant at the maximum flow rate that can be achieved with the selected liner size. This flow rate can be identified from the manufacturer's specifications table such as that shown in Table 3.6. If the table is not available for a given pump horsepower rating P_{Hp}, this maximum rate can be calculated by

$$q_{max} = \frac{1{,}714 P_{Hp} E}{p_{max}} \tag{4.12}$$

where

q_{max} = maximum flow rate for the liner, psi or kPa
E = pump efficiency, dimensionless
p_{max} = maximum allowable pump pressure for the liner, psi or kPa

This flow rate q_{max} should be used until a critical depth is reached at which $\Delta p_d = \frac{p_p}{m+1}$ if the maximum bit hydraulic power criterion is used or $\Delta p_d = \frac{2p_p}{m+2}$ if the maximum jet impact force criterion is used. Before this critical depth is reached, the parasitic pressure loss can be calculated using

$$\Delta p_d = c q_{max}^m \tag{4.13}$$

and the pressure differential available for the bit is

$$\Delta p_b = p_{max} - c q_{max}^m \tag{4.14}$$

The total flow area of nozzles should be determined by

$$A_T = \frac{0.00912 q_{\max}}{C_d}\sqrt{\frac{\rho}{\Delta p_b}} \tag{4.15}$$

After the critical depth is reached, the flow rate is decreased with a subsequent increase in depth to maintain $\Delta p_d = \frac{p_p}{m+1}$ if the maximum bit hydraulic power criterion is used or $\Delta p_d = \frac{2p_p}{m+2}$ if the maximum jet impact force criterion is used. The flow rate at the depth of interest should be set to be

$$q = \sqrt[m]{\frac{\Delta p_d}{c}} \tag{4.16}$$

where

$$\Delta p_d = \frac{p_p}{m+1} \tag{4.17}$$

for the maximum bit hydraulic power criterion, and

$$\Delta p_d = \frac{2p_p}{m+2} \tag{4.18}$$

for the maximum jet impact force criterion. The total nozzle area should be determined to be

$$A_T = \frac{0.00912 q}{C_d}\sqrt{\frac{\rho}{\Delta p_b}} \tag{4.19}$$

where

$$\Delta p_b = \frac{m}{m+1} p_p \tag{4.20}$$

according to the hydraulic horsepower criterion, and

$$\Delta p_b = \frac{m}{m+2} p_p \tag{4.21}$$

according to the maximum jet impact force criterion.

The flow rate, however, should never be reduced below the minimum flow rate q_{min} required to lift cuttings. The minimum flow rate q_{min} should be maintained, and larger nozzles should be used in the subsequent depth to maintain pump pressures less than p_{max}. Nozzles should then be sized not honoring the optimum pressure ratio as follows. First, calculate the parasitic pressure losses using

$$\Delta p_d = cq_{min}^m \tag{4.22}$$

Second, the allowable pressure drop at the bit is calculated by

$$\Delta p_b = p_{max} - \Delta p_d \tag{4.23}$$

Finally, the total nozzle area is calculated by

$$A_T = \frac{0.00912 q_{min}}{C_d} \sqrt{\frac{\rho}{\Delta p_b}} \tag{4.24}$$

If the maximum nozzle velocity criterion is employed, Eq. (2.87) indicates that the nozzle velocity is maximum when the pressure drop at the bit is maximum. To maximize the pressure drop at the bit, the parasitic pressure losses need to be minimized. For given well geometries and fluid properties, the parasitic pressure can be minimized with the minimum flow rate corresponding to the minimum annulus velocity required to lift cuttings. Based on this theory, bit nozzles are sized as follows.

First, calculate the parasitic pressure losses using

$$\Delta p_d = cq_{min}^m \tag{4.25}$$

Second, the allowable pressure drop at the bit is calculated by

$$\Delta p_b = p_{max} - \Delta p_d \tag{4.26}$$

Finally, the total nozzle area is set to be

$$A_t = \frac{0.00912 q_{min}}{C_d} \sqrt{\frac{\rho}{\Delta p_b}} \tag{4.27}$$

4.3.3 Application Examples

This section illustrates the applications of hydraulics models in drilling hydraulics design and calculations of surge, swab, and critical running speeds using commercial software.

Illustrative Example 4.2

General data are given as follows for drilling hydraulics analysis:

Surface equipment: Type III
Casing size: 8.535 in, bottom at 8,200 ft
Hole size: 8.500 in, bottom at 11,500 ft
Drill pipe: 5 × 4.276 in
Drill collar: 6.25 × 2.813 in, 656 ft
Nozzle coefficient: 0.95
Nozzle size: 3 × 9/32 in
Bit size: 7.875 in
Flow rate: 285 gpm

Weak zone

Measured depth: 11,500 ft
Pore pressure gradient: 9.2 ppg
Fracture pressure gradient: 10.65 ppg
Mud weight: 10 ppg, Bingham plastic fluid
Plastic viscosity: 23 cp
Yield point: 14.79 lb/100 ft^2
Maximum pump pressure: 3,200 psi
Maximum pump horsepower: 1,184 hp
Minimum annular velocity: 82 ft/min

Cuttings properties

Rate of penetration: 33 ft/hr
Cuttings diameter: 0.197 in
Cuttings density: 21 ppg

The goal of the drilling hydraulics analysis is to make sure the following concerns of drilling engineers are addressed:

1. The required pump pressure can be delivered by the selected pump.
2. The downhole equivalent circulating density (ECD) is between the pore and fracture gradient so there will be no kick or loss of circulation.
3. The cuttings concentrate is controlled. It is normally required that the cuttings concentrate should be less than 5%.
4. Under the condition of meeting other requirements, the flow rate should be chosen so either bit hydraulic horsepower or jet impact force is maximized.

Table 4.1 shows an output of HYDPRO, which is a drilling software model developed by Pegasus Vertex. It indicates that the required pump pressure is 2,910 psi, which is less than the maximum pump pressure of 3,200 psi. The pressure profile in the system is presented in Figure 4.1. The computed cuttings concentrate profile is shown in Figure 4.2, which indicates that the maximum cuttings concentration, 0.706%, is lower than the normally permissible value of 5%.

Table 4.1 Output of Computer Software HYDPRO

#	Parameter	Value	Unit
1	Bit MD/TVD	11500./11500.	(ft)
2	Flow rate	285	(gpm)
3	Pump pressure	2910	(psi)
4	Pump output horsepower	484.	(HP)
5	Pressure loss: surface/pipe	15/433	(psi)
6	Pressure loss: motor/nozzle	0/2153	(psi)
7	Pressure loss: annulus	309	(psi)
8	Pressure loss: total	2910	(psi)
9	Nozzle hydraulic power	358.	(HP)
10	Percent power through nozzle	74.0	(%)
11	Jet impact force	724	(lbf)
12	Bit HHP per sq. in of bit size	7.35	(hp/in2)
13	Bit nozzle velocity	491.	(ft/s)
14	Volume: hole/pipe displacement	811.9/90.6	(bbl)
15	Volume: inside pipe/annulus	197.7/523.6	(bbl)
16	Circulation time: down/up	29.13/77.17	(min)
17	Circulation time: full cycle	106.30	(min)
18	Weak zone: MD/TVD	10000./10000.	(ft)
19	Pore/frac gradi. @ weak zone	9.00/11.50	(ppg)
20	Pore/frac press. @ weak zone	4680/5980	(psi)
21	ECD @ weak zone	10.49	(ppg)
22	Circulation pressure @ weak zone	5455	(psi)
23	Initiate flow: ECD @ weak zone	10.54	(ppg)
24	Initiate flow: pressure @ weak zone	5483	(psi)
25	Initiate flow: pump pressure	614	(psi)
26	Minimum annular velocity	146.0	(ft/min)
27	Minimum cuttings velocity	116.8	(ft/min)
28	Bit hydraulics optimization:		
29	Maximum pump pressure	3200	(psi)
30	Maximum pump output HP	1184.	(HP)
31	Mini/maxi flow rates	160/571	(gpm)
32	Maximum bit hydraulics HP:		
33	Optimum flow rate	432	(gpm)
34	Optimum pressure drop @ bit	2036	(psi)
35	% total system press loss @ bit	64	(%)
36	Bit hydraulic power	513.	(HP)
37	Jet impact force	1067	(lbf)
38	Bit HHP per sq. in of bit size	10.53	(hp/in2)
39	Recommended TFA	0.290	(in2)
40	Recommended nozzles	11 11 11	(1/32in)
41	Maximum jet impact force:		
42	Optimum flow rate	537	(gpm)
43	Optimum pressure drop @ bit	1493	(psi)
44	% total system press loss @ bit	47	(%)
45	Bit hydraulic power	468.	(HP)
46	Jet impact force	1137	(lbf)
47	Bit HHP per sq. in of bit size	9.61	(hp/in2)
48	Recommended TFA	0.422	(in2)
49	Recommended nozzles	13 14 14	(1/32in)

(*Continued*)

Illustrative Example 4.2 *(Continued)*

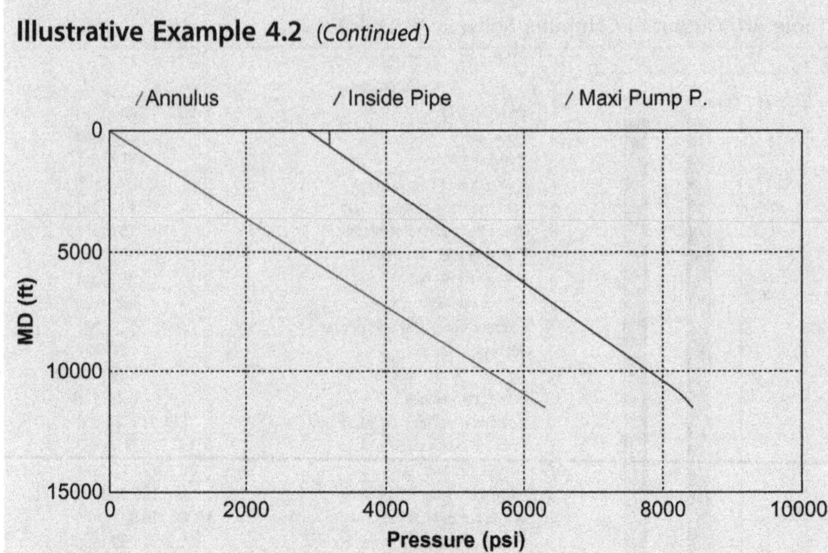

Figure 4.1 Pressure profile in the circulating system.

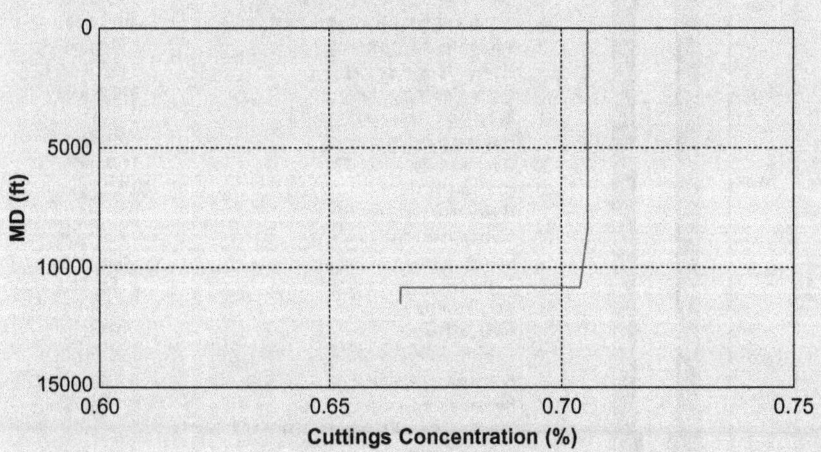

Figure 4.2 Cuttings concentration profile in the circulating system.

The calculated effects of the mud flow rate on other parameters are shown in Table 4.2. These parameters include hydraulic horsepower, jet impact force, ECD, pump pressure, and pressure components. The calculated hydraulic horsepower profile is illustrated in Figure 4.3. It shows

Table 4.2 Mud Flow Rate-Related Parameter Values Given by HYDPRO

	A	B	C
1	Hydraulics analysis at string depth	11500.	(ft)
2	Allowable maximum flow rate	570	(gpm)
3	Volume inside pipe	197.7	(bbl)
4	Volume inside annulus	523.6	(bbl)
5	Volume inside pipe and annulus	721.3	(bbl)
6	Weak zone MD	11500.	(ft)
7	Weak zone TVD	11500.	(ft)
8	Fracture pressure gradient @ weak zone	10.65	(ppg)
9	Fracture pressure @ weak zone	6369	(psi)
10	Pore pressure gradient @ weak zone	9.20	(ppg)
11	Pore pressure @ weak zone	5502	(psi)
12	Current flow rate	285	(gpm)
13	Pump pressure	2910	(psi)
14	Pump horsepower	484.	(HP)
15	Circulation pressure @ weak zone	6289	(psi)
16	ECD @ weak zone	10.52	(ppg)
17	Pressure loss: surface equip.	15	(psi)
18	Pressure loss: inside pipe	433	(psi)
19	Pressure loss: motor, etc	0	(psi)
20	Pressure loss: nozzle	2153	(psi)
21	Pressure loss: annulus	309	(psi)
22	Pressure loss: total	2910	(psi)
23	Circulation time: down	29.13	(min)
24	Circulation time: up	77.17	(min)
25	Circulation time: full cycle	106.30	(min)

	Flow rate (gpm)	P. @ weak zone (psi)	ECD @ weak zone (ppg)	Pump P. (psi)	P. drop_pipe (psi)	P. drop_annulus (psi)	P. drop_bit (psi)
1	200	6271	10.49	1638	287	291	1060
2	250	6281	10.50	2315	357	301	1657
3	300	6292	10.52	3188	490	312	2386
4	350	6302	10.54	4211	642	322	3247
5	400	6313	10.56	5385	811	333	4242
6	450	6327	10.58	6713	997	347	5368
7	500	6344	10.61	8192	1201	364	6627
8	550	6361	10.64	9821	1420	381	8019
9	600	6391	10.69	11610	1656	411	9543
10	650	6451	10.79	13579	1907	471	11200
11	700	6515	10.90	15699	2174	535	12990
12	750	6583	11.01	17971	2457	603	14912
13	800	6654	11.13	20394	2754	674	16966

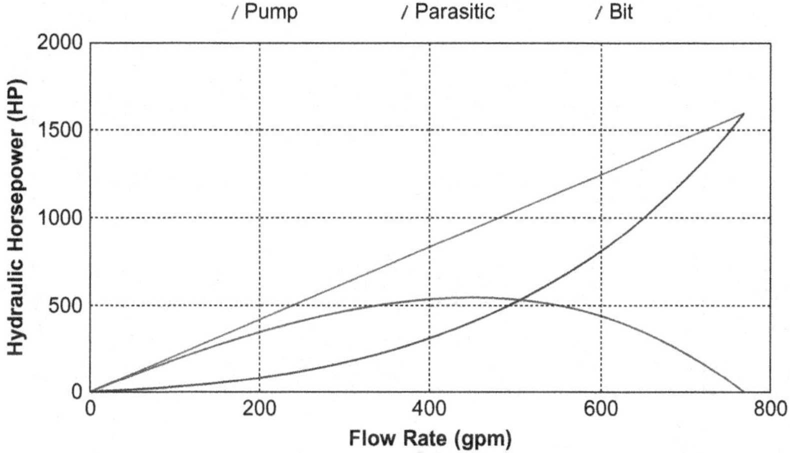

Figure 4.3 Hydraulic horsepower versus flow rate profile in the circulating system.

that the bit hydraulic horsepower will reach 50% of pump horsepower at a mud flow rate of about 500 gpm.

The calculated jet impact force profile is presented in Figure 4.4, which indicates that the impact force will be maximal at a mud flow rate of about 570 gpm. The calculated equivalent circulating density profile is illustrated in Figure 4.5. It shows that the ECD profile is between the pore pressure gradient, 9.2 ppg, and the fracture pressure gradient, 10.65 ppg, if the mud flow rate is less than 560 gpm—that is, neither

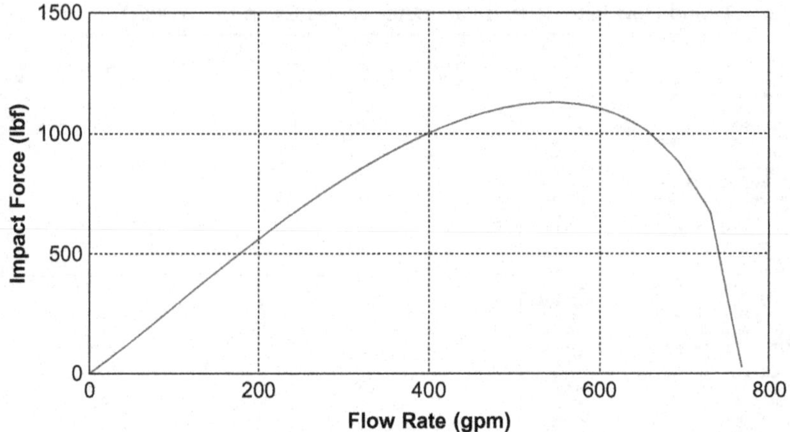

Figure 4.4 Jet impact force versus flow rate profile in the circulating system.

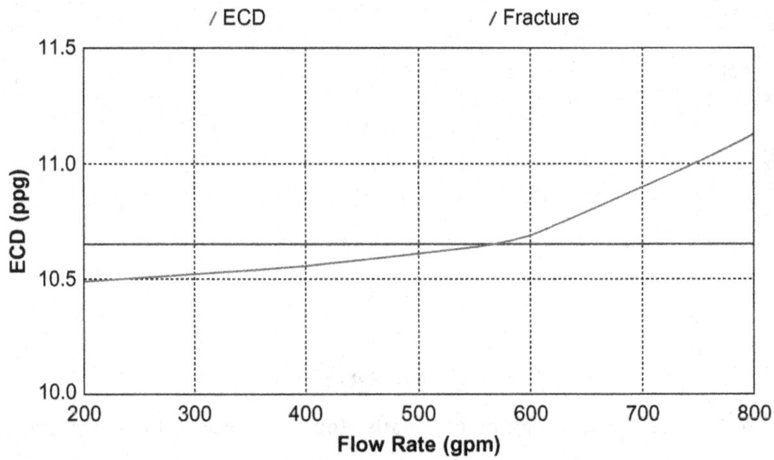

Figure 4.5 ECD at 11,500 (ft) versus flow rate profile in the circulating system.

kick nor loss of circulation is expected at mud flow rate values lower than 560 gpm.

The calculated pump pressure profile is plotted in Figure 4.6. It indicates that the maximum allowable flow rate is 300 gpm for the pump in order to maintain the pump pressure below the maximum allowable value of 3,000 psi. The calculated pressure component profile is presented in Figure 4.7. It shows that the pressure drop at the bit is always the major component.

When a drilling string or casing is moved, it displaces the mud in the hole, which leads to pressure variations. A pressure increase due to a downward pipe movement is called surge pressure, whereas a pressure decrease due to an upward pipe movement is called swab pressure. Excessive swab pressure may initiate a kick, while surge pressure is detrimental in that it frequently is large enough to fracture a formation. This is particularly true for ERD wells, slim holes, and deepwater offshore

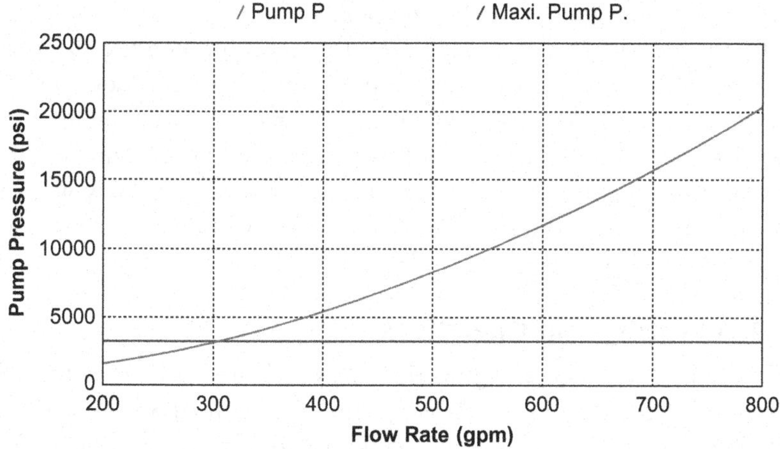

Figure 4.6 Pump pressure versus mud flow rate.

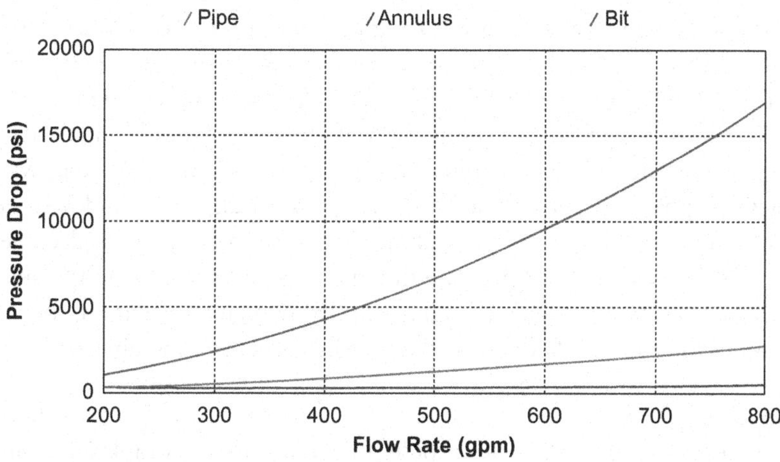

Figure 4.7 Pressure components versus mud flow rate.

drilling because of restricted flow paths and the limited number of casings and liners. Accurately predicting surge and swab pressures is of great importance in wells, where the pressure must be maintained within narrow limits to ensure trouble-free drilling and completion operations.

New drilling and completion technologies are challenging many aspects of our operations. For example, running liners in a subsea casing string with very tight tolerance can cause extremely high surge pressures. Autofill float equipment and other new tools such as flow diverters (also called circulation subs) have been developed to reduce surge pressure, and they are effective. The questions are, what will the surge/swab pressures be and what are the optimal tripping speeds?

To thoroughly analyze surge pressure, a comprehensive surge and swab hydraulics computer model, SurgeMOD, has been developed to assist in the analysis and design of tripping operations, especially for deepwater wells or wells using new tools such as autofill float equipment, circulation subs, and so on. The program simulates fairly complex wellbore configurations, including multiple pipe sizes, wellbore intervals, and annular sections, with very tight tolerance.

4.3.4 Analyzing Trip Operations

This section discusses the engineering analyses behind trip operations for different pipe end conditions: closed, open, open with autofill or a bit, and with a flow diverter. These four conditions are illustrated in Figure 4.8. We will discuss the controlling parameters that affect surge pressure using SurgeMOD. The surge and swab pressure analysis has two components: to predict surge and swab pressure for a given running speed (analysis mode) and to calculate optimal trip speeds at different string depths without breaking down formations or causing a kick at weak zones (design mode).

As pipe is moved downward into a well, the original mud is displaced by the new volume of the extending pipe, and the mud must move upward. When the pipe is closed or contains a float sub, all displaced fluid passes up the annulus. The flow rate in the annulus is equal to the pipe displacement rate. It is therefore easy to calculate the frictional pressure drop in the annulus. Surge pressure is calculated using standard hydraulics equations, but the equations must be modified to account for movement of the pipe wall.

Fully Open Pipe

If the pipe is open-ended, the problem becomes more complicated, since the distribution of flow between the inside pipe and the annulus cannot

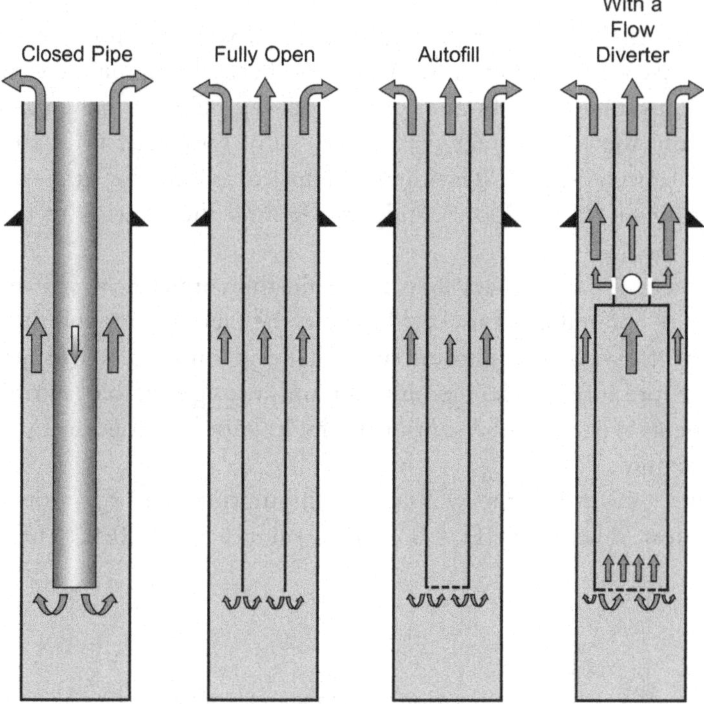

Figure 4.8 Different pipe end conditions.

be determined by any simple method. A split of flow going to the annulus and pipe interior is iteratively calculated. A numerical method must be used to make sure that the sum of the resulting frictional pressure drops inside the pipes is equal to that of all the annular sections.

Autofill or Pipe with a Nozzle

The difference between autofill float equipment and a fully open pipe is the additional pressure drop across the orifices on the autofill float equipment. Depending on the total flow area of the autofill equipment, the resulting surge pressure can vary significantly. The actual surge pressure should be between the pressure of the closed pipe and that of fully open pipe.

Pipe with a Flow Diverter

A new tool, commonly referred to as flow diverter valve or circulation sub, can be used in conjunction with autofill float equipment. This tool, which is located on the drill pipe immediately above the liner, has ports open to the drill pipe annulus. These ports allow the fluid trapped in the liner to escape from the narrow drill pipe interior to the larger annulus between the drill pipe and casing. Equipped with this tool, the system now has two areas of fluid communication between the pipe interior and the annulus: one at the bottom of the liner (autofill float equipment) and one at the top of the liner (circulation sub). Displaced fluid seeks the path of least resistance. This device helps to reduce the surge pressure depending on the wellbore configurations and the location of the flow diverter tool.

Numerical methods are used to obtain the correct flow split percentage when communications exist between the pipe interior and the annulus. The flow split is chosen such that the sum of hydrostatic and frictional pressures in the pipe interior and through the bit (autofill float equipment) will equal the sum of the hydrostatic and frictional pressures in the annulus.

Figure 4.9 shows the wellbore configuration used for an example calculation. The riser (ID = 17.755 in) depth is 3,500 ft. The casing

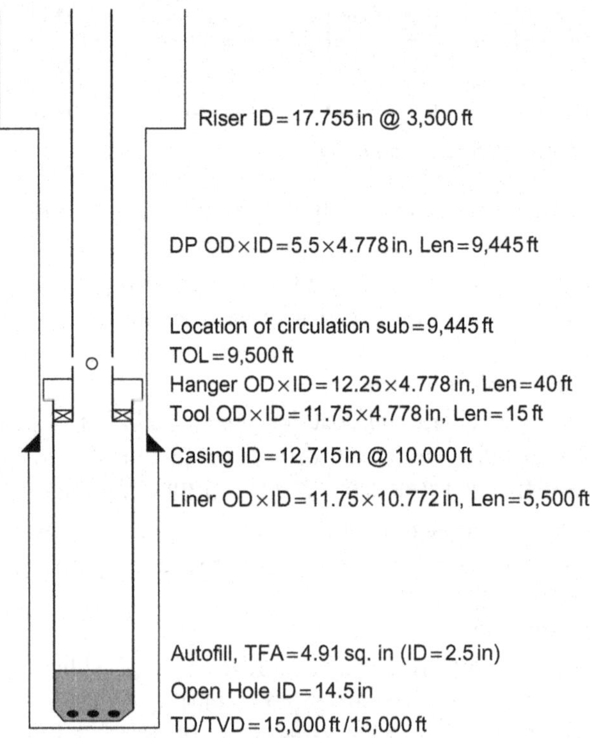

Figure 4.9 Example of a wellbore configuration.

(ID = 12.715 in) was set at the depth of 10,000 ft. The open hole has a diameter of 14.5 in. The total depth of this vertical well is 15,000 ft. The mud weight is 11 ppg, with a plastic viscosity of 20 cp and a yield point of 15 lb/100 ft^2. The weak zone is at 15,000 ft, with a pore and fracture pressure gradient of 10.6 ppg and 11.5 ppg, respectively.

The goal is to run 5,500 ft of liner (11¾-in OD, 60 lb/ft, 10.772-in ID) to the bottom. The autofill float equipment has an orifice with a total flow area (TFA) of 4.91 sq. in. The challenge is to run the liner through the casing and the open hole section without fracturing the formation. Note that the annular radial clearance between the casing and the liner is 0.4825 in.

We will first use SurgeMOD to calculate the surge and swab pressures for a given tripping speed. Figure 4.10 shows the simulated bottomhole equivalent mud weight (EMW) versus the string depth during the trip-in operation at 50 ft/min. If the pipe is closed at the end, the loss of circulation would occur when the liner reaches 5,000 ft. A liner with autofill float equipment (TFA = 4.91 sq. in) could be run to TD without fracturing the formation. If a circulation sub is placed above the liner, this trip speed would be safe for the entire wellbore as well. EMWs for both closed and open pipe decrease after the pipe passes 10,000 ft due to the larger diameter of the open hole section. We can see that a fixed tripping speed of 50 ft/min may be safe at total depth (TD), but it would have already fractured the formation before it reaches TD for closed pipe.

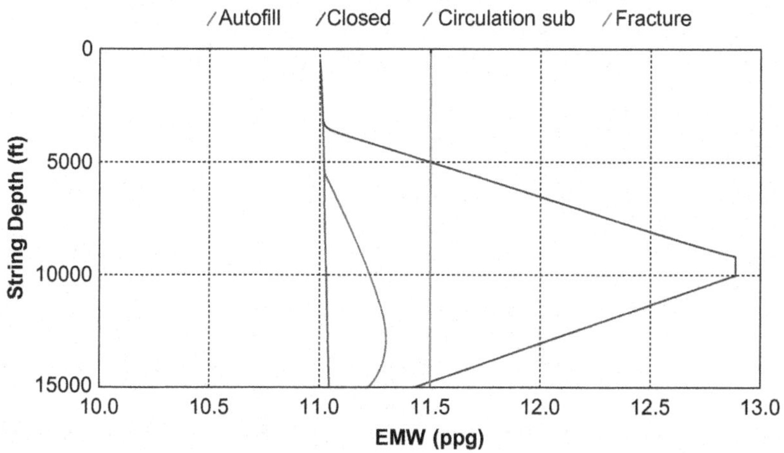

Figure 4.10 Calculated bottomhole EMW for trip-in operation. Surge EMW @ 15,000 (ft); spd = 50.0 (ft/min).

Figure 4.11 presents the calculated bottomhole equivalent mud weight for a trip-out operation. The maximum pipe running speed without fracturing the formation or causing a kick at a weak zone is analyzed as follows. Figure 4.12 shows the maximum tripping speeds for different

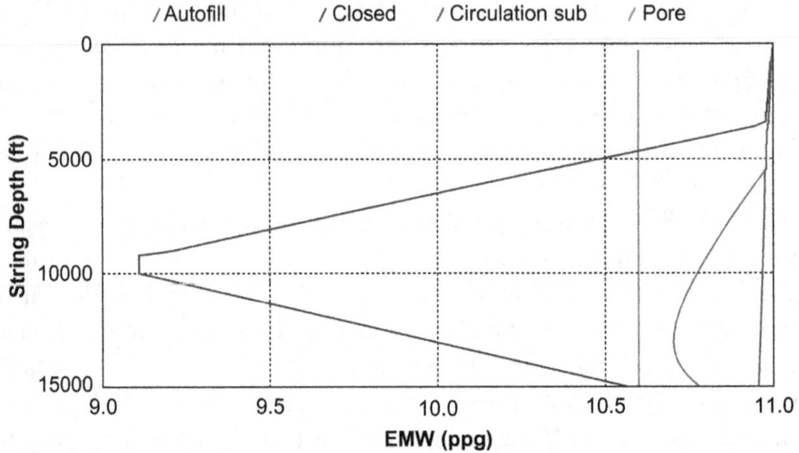

Figure 4.11 Calculated bottomhole EMW for trip-out operation. Swab EMW @ 15,000 (ft); spd = 50.0 (ft/min).

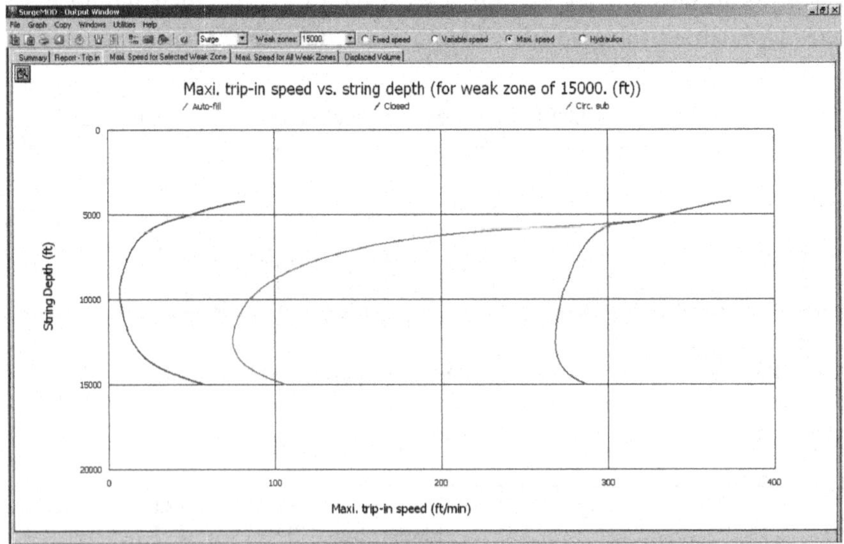

Figure 4.12 Calculated maximum allowable trip-in speed for a weak zone of 15,000 (ft).

pipe end conditions at various depths. As we can see from this graph, engineers must pay close attention before the pipe reaches 10,000 ft. At 10,000 ft, the narrowest annular section is the longest, thus producing the maximum surge pressure. When the liner enters the larger open hole section, trip speeds can be increased. The most dangerous string depth is not necessarily at the bottom of the well. Note also that the curves for a pipe with autofill float equipment and a pipe with a circulation sub coincide above the string depth of 5,555 ft. This is because above that depth, the circulation sub, which is located at a depth of 9,445 ft, is not in the well.

Figure 4.13 shows the sensitivities of surge pressures to tripping speeds for pipe ending conditions that involve pipes that are closed, have autofill float equipment, and have a flow diverter. The SurgeMOD program is equipped with pipe tripping animation so the user can view the positions of the pipe and pressure variation simultaneously. Figure 4.13 is for the sensitivity of surge pressures at a depth of 15,000 ft. If we increase the tripping speed, the surge pressure for the closed pipe will increase sharply and the surge pressure for the pipe with a circulation sub will gradually increase. The curve for the pipe with the autofill float equipment lies between them.

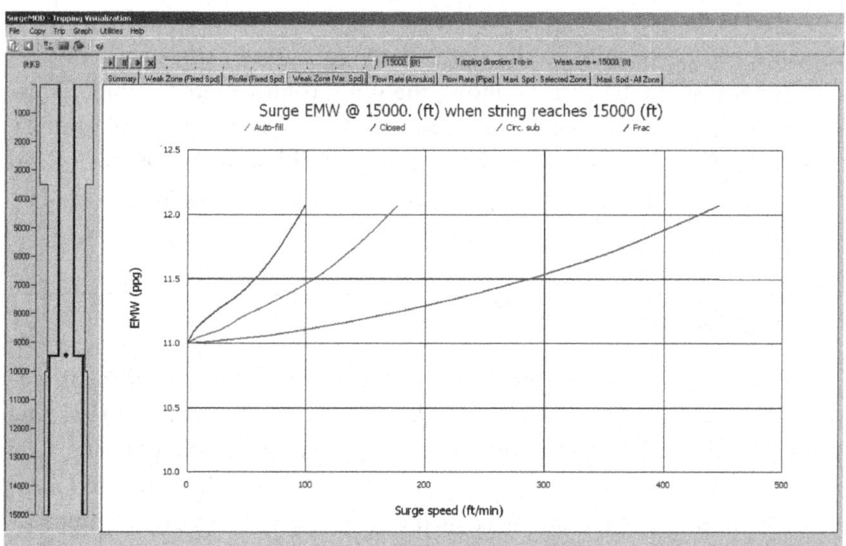

Figure 4.13 Sensitivities of surge pressures to tripping speeds. Surge EMW @ 15,000 (ft) when string reaches 15,000 (ft).

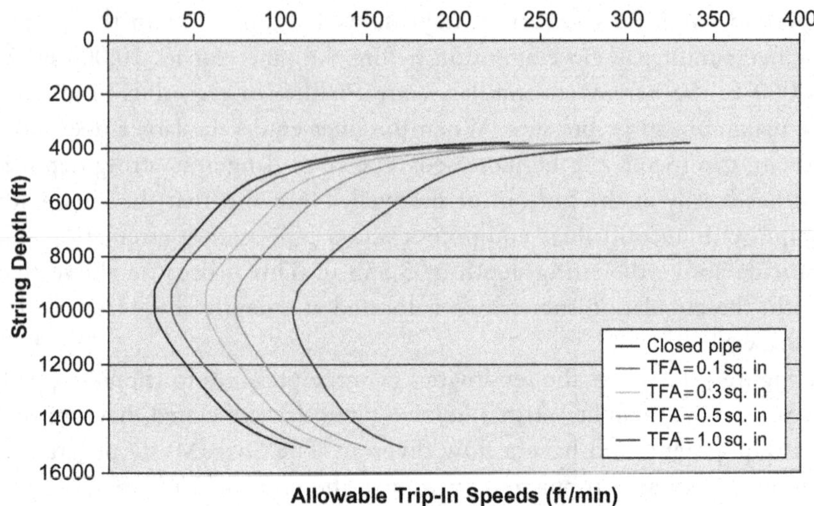

Figure 4.14 Effects of TFA of autofill float equipment.

We have seen that the pipe ending conditions greatly affect surge pressures. Now we will see how the total flow area of autofill float equipment affects the surge pressure for open-ended pipes. Figure 4.14 shows the calculated maximum trip-in speeds versus the string depths for various total flow areas. As the TFA of autofill float equipment increases, the optimal trip-in speed curve shifts to the right, allowing a greater trip-in speed. This occurs because the large TFA allows the fluid to move into the pipe interior more freely, reducing the amount of fluid flowing into the annulus. This redistribution of flow reduces the overall frictional pressure drop along the flow paths inside the pipe and outside in the annulus.

SUMMARY

This chapter presented some background and application procedures for optimizing drilling hydraulics programs. Accurately predicting parasitic pressure loss is essential to hydraulics program design. Using field measurements can significantly improve drilling hydraulics on site. Computer software provides an efficient means of optimizing drilling hydraulics programs.

REFERENCES

Kendall, W.A., Goins, W.C., 1960. Design and operation of jet bit programs for maximum hydraulics horsepower, impact force, or jet velocity. Trans. AIME 219, 238–247.

Speer, J.W., 1958. A method for determining optimum drilling techniques. Drill. Prod. Prac. API, 130–147.

Sutko, A.A., Myers, G.M., 1970. The effect of nozzle size, number, and extension on the pressure distribution under a tricone bit. SPE Paper 3109, presented at the Fall Meeting of SPE of AIME, October, Houston.

PROBLEMS

4.1 Predict the parasitic pressure loss under the following conditions:
Total depth: 9,950 ft (3,036 m)
Casing: 9⅝ in, 43.5 lb/ft (8.755-in ID), set at 6,500 ft (1,982 m)
Open hole: 7⅞ in from 6,500 ft to 9,950 ft
Drill pipe: 9,500 ft of 4½-in, 16.6 lb/ft (3.826-in ID)
Drill collar: 450 ft of 6¾-in OD and 2¼-in ID
Surface equipment: Combination 4
Mud weight: 12.5 ppg (1,498 kg/m^3)
Plastic viscosity: 40 cp (0.04 Pa-s)
Yield point: 15 lb/100 ft^2 (100 N/m^2)
Mud flow rate: 350 gpm (1.33 m^3/min)

4.2 For the data in Problem 4.1, select a liner size for two TSC WF700 Triplex pumps. Assume the flow rate exponent m = 1.75 and the maximum bit hydraulic horsepower criterion. Additional data are given as follows:
Cuttings specific gravity: 2.65 water = 1
Particle sphericity: 0.80 ball = 1
Rate of penetration: 70 ft/hr (21.3 m/hr)
Rotary speed: 60 rpm
Cuttings concentration: 8%

4.3 For the data in Problems 4.1 and 4.2, design the mud flow rates and the bit nozzle sizes at depths from 6,500 ft (1,981 m) to 9,950 ft (3,033 m).

4.4 Using the data given in Problem 4.1 and the maximum bit hydraulic power criterion, calculate the required pump pressure and select a pump from Table 3.6.

4.5 Using the data given in Problem 4.1 and the maximum bit hydraulic jet impact force criterion, calculate the required pump pressure and select a pump from Table 3.6.

PART TWO

Gas Drilling Systems

Different types of gases are used as drilling fluids to drill geotechnical boreholes, mining boreholes, oil and gas recovery wells, and water wells. Since gases have considerably lower densities than water, they are used for drilling relatively dry rock formations that do not produce a significant amount of liquids during drilling. The main purpose of gas drilling is to improve drilling performance by increasing the rate of penetration in hard formations.

This part of the book provides drilling engineers with basic information about gas drilling systems and techniques that are used to reduce the risks associated with gas drilling to safely achieve the best drilling performance. This part consists of three chapters:

Chapter 5: Equipment in Gas Drilling Systems
Chapter 6: Gas Compressors
Chapter 7: Safe Gas Drilling Operations

CHAPTER FIVE

Equipment in Gas Drilling Systems

Contents

5.1 INTRODUCTION

The gases used in gas drilling can be air, natural gas, or nitrogen. Gas injection systems are similar when different types of gases are used. Figure 5.1 shows a typical gas drilling system that uses air (Lyons et al., 2009). The air travels (1) from the atmosphere to the compressor, (2) from the compressor through the standpipe and the kelly pipe to the drill string that consists of drill pipes and drill collars, (3) through the drill string to the bit, (4) through the bit and up the annular space between the drill string and the borehole (open hole and cased hole sections) to the surface, and (5) through the rotating head to the blooey line and back to the atmosphere. This chapter provides a brief introduction to the equipment that controls gas injection pressure.

5.2 SURFACE EQUIPMENT

Figures 5.2, 5.3, and 5.4 show layouts of the surface equipment in air drilling systems, natural gas drilling systems, and nitrogen drilling systems, respectively. In all of these systems, the most important equipment includes the compressors and the boosters (high-pressure, high-volume compressors). Different types of gas compressors are used in the oil and gas industry. These designs vary greatly in volume and pressure ratings. Gas compressors are very similar to liquid pumps in their basic design and operation. The basic difference is that compressors move compressible fluids, whereas pumps move incompressible fluids.

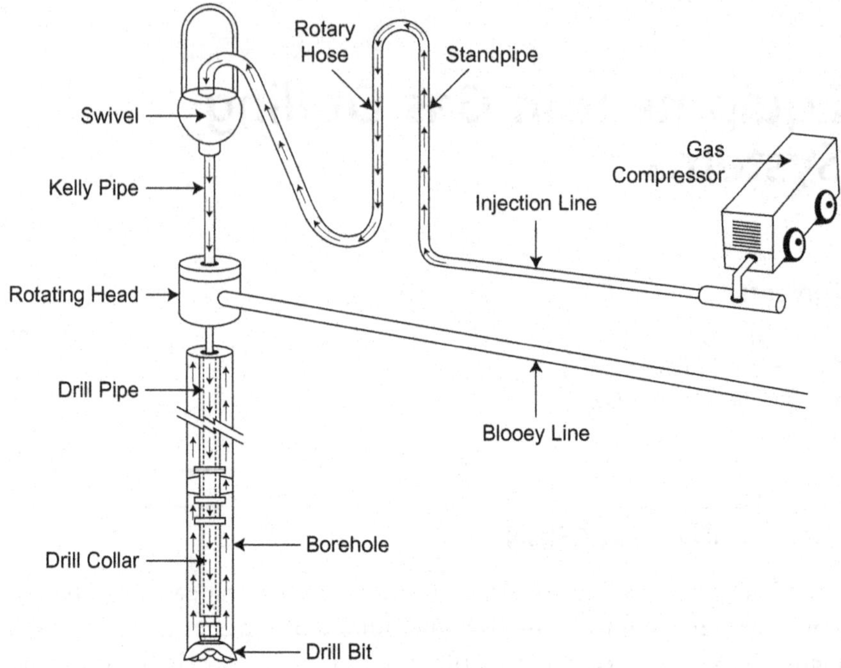

Figure 5.1 A typical gas injection system.

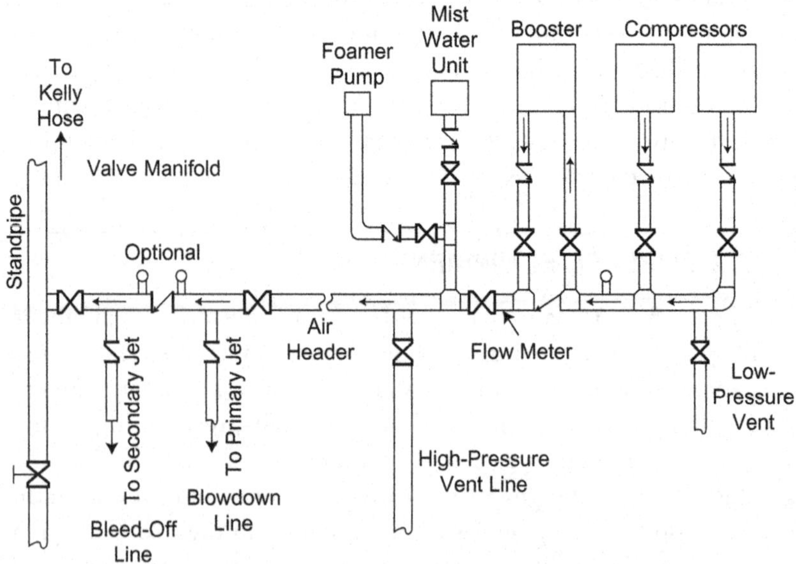

Figure 5.2 A layout of the surface equipment in an air drilling system.

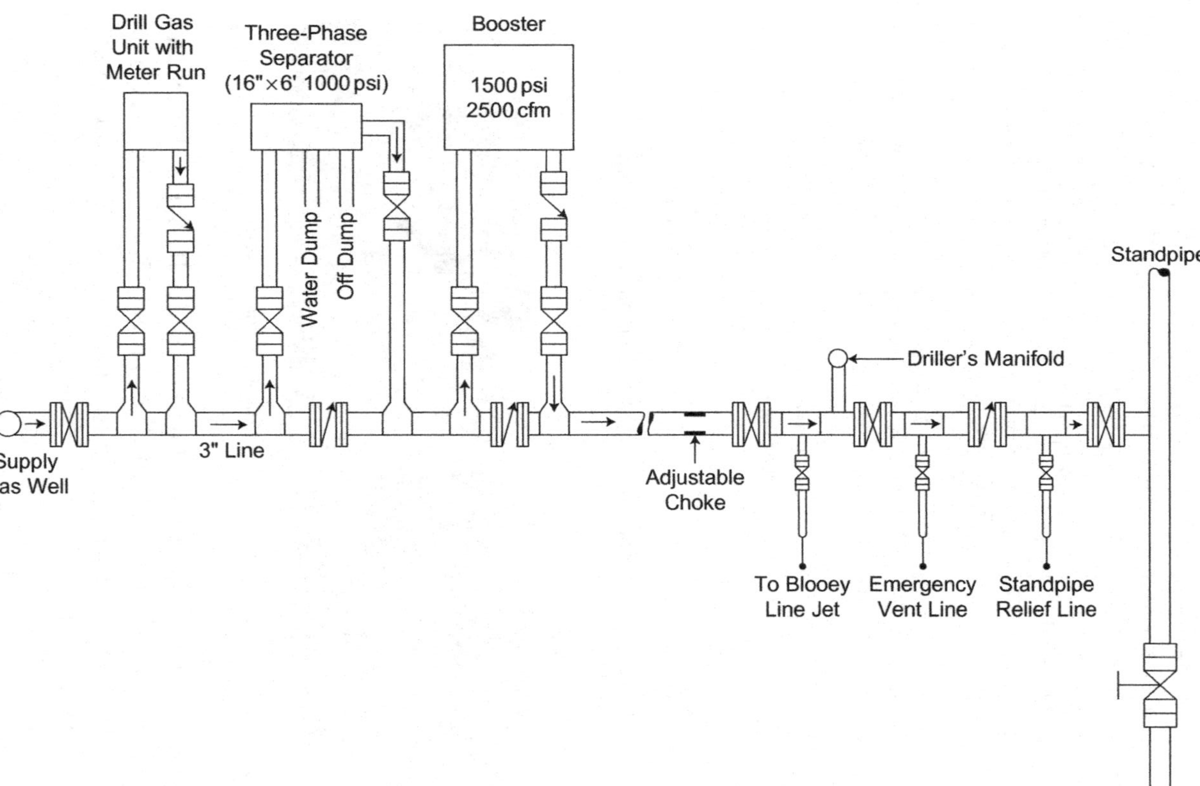

Figure 5.3 A layout of the surface equipment in a natural gas drilling system.

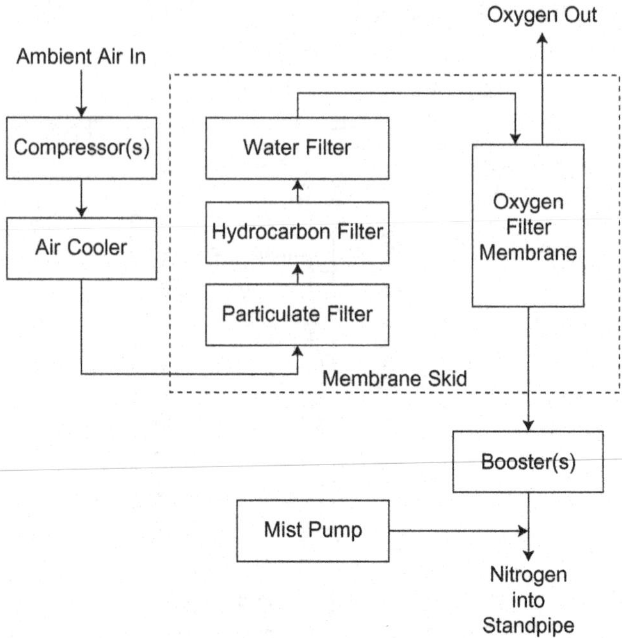

Figure 5.4 A layout of the surface equipment in a nitrogen gas drilling system.

Figure 5.5 Air Drilling Specialties IR-XHP1170FSCAT compressor.

Figure 5.5 shows the Air Drilling Specialties IR-XHP1170FSCAT compressor with an aftercooler. This compressor is a rotary screw type (Lyons et al., 2001) that delivers air at a flow rate of 1,170 cfm (33 m^3/min) in a pressure range of 150 to 375 psig (10 to 26 bar).

The blooey line (Figure 5.6) serves as a returning path for the gas stream that contains rock cuttings and a small amount of fluids from the formations (Hook et al., 1977). A pilot light is installed at the end of the line to burn any natural gas from the returning gas stream. The blooey line is equipped with primary and secondary jets to allow the safe venting of the top of the wellhead when the well is producing natural gas or other dangerous gases. They are also useful for cleaning the drill cuttings in the blooey line.

Another piece of equipment at the surface is the rotating head, or diverter, which is installed upstream of the blooey line (Figure 5.7).

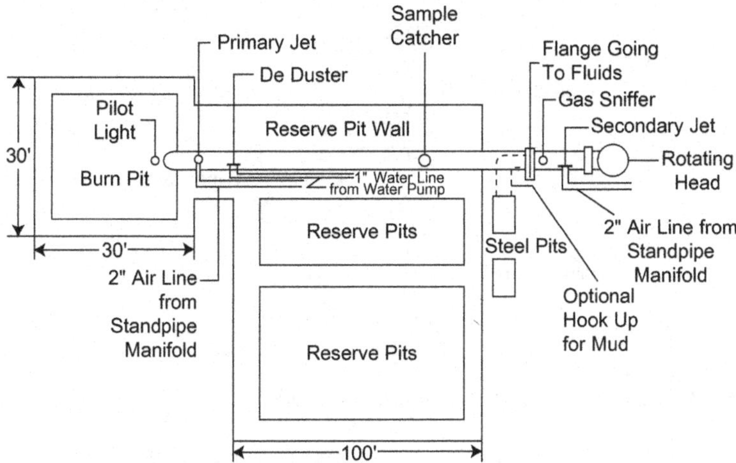

Figure 5.6 A blooey line layout in a gas drilling system.

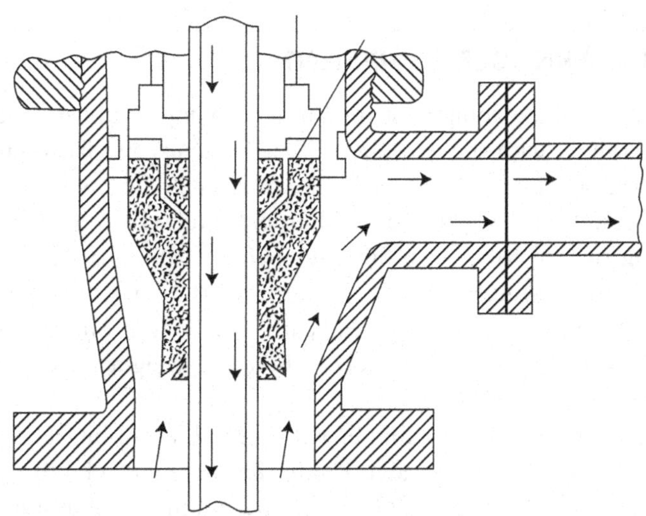

Figure 5.7 A rotating head in a gas drilling system.

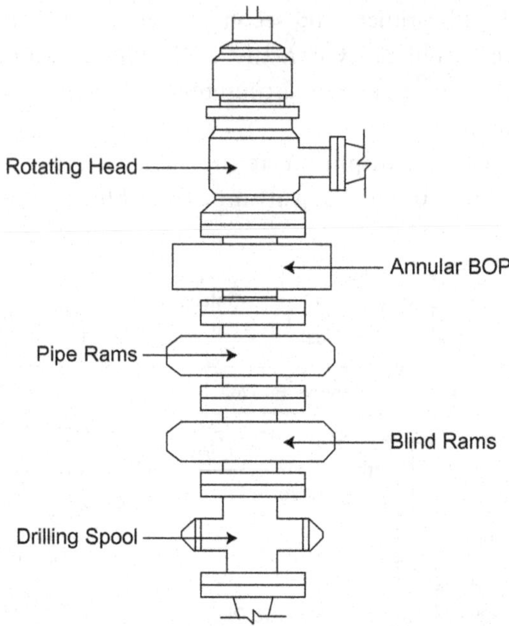

Figure 5.8 A typical BOP stack with a rotating head.

It diverts the returning gas stream to the blooey line so normal operations at the drilling floor can be performed safely. The rotating head is installed on the top of the blowout preventor (BOP) stack (Figure 5.8).

5.3 DOWNHOLE EQUIPMENT

The downhole equipment used in gas drilling is similar to that used in liquid drilling, with the exception of air hammers and flat-bottom bits. Using a combination of these tools allows faster drilling with less weight on bit. These tools are used for drilling extremely hard formations and for drilling surface rocks in mountains before drill collars are added to obtain adequate weight on bit. Sometimes they are utilized to drill straight holes in crooked hole areas with low weight on bit. They can also be used to drill horizontal holes where the weight on bit is limited due to high torque and drag.

Figure 5.9 shows an air hammer with a flat-bottom bit. The gas flow path is self-controlled by a valve inside the hammer, which moves the piston to produce 1,200 to 1,800 strokes per minute. The piston hits the top of the bit and generates enough force on the bottomhole that the bit

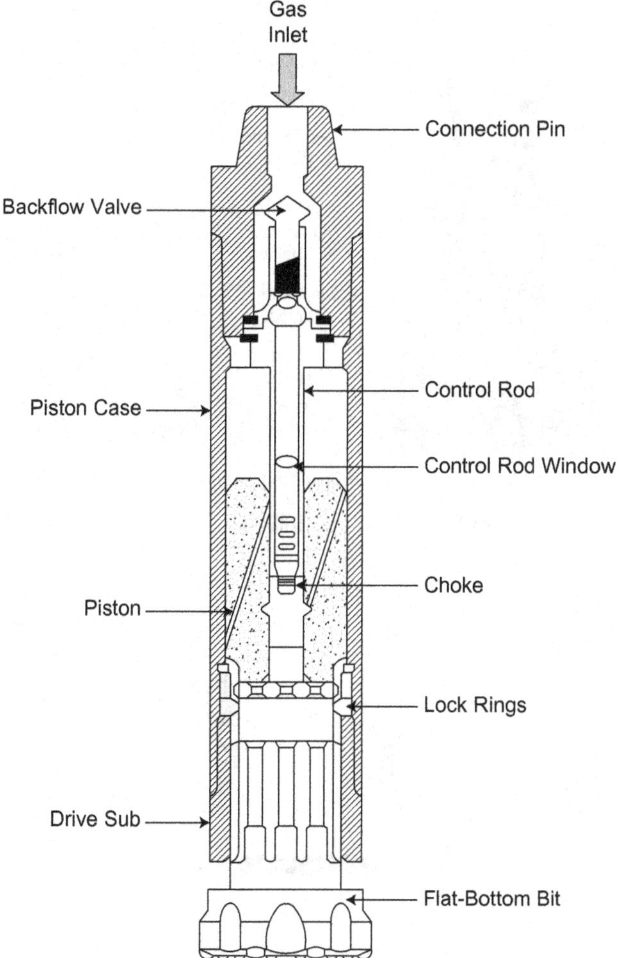

Figure 5.9 An air hammer with a flat-bottom bit.

can cut the formation rock without the need for a heavy weight on bit. A few hundred psi of pressure drop across the air hammer is required under drilling conditions.

A flat-bottom bit is shown in Figure 5.10. Diamond-enhanced inserts are installed on the flat-bottom face of the bit. The rotational drive spines allow the bit to rotate while cutting rocks forward. This avoids repeated cutting actions on the same points of the bottomhole.

In addition to air hammers and flat-bottom bits, float valves are widely used in gas drilling (Figure 5.11). They are used to prevent backflow of gas up the drill string (Lyons et al., 2009).

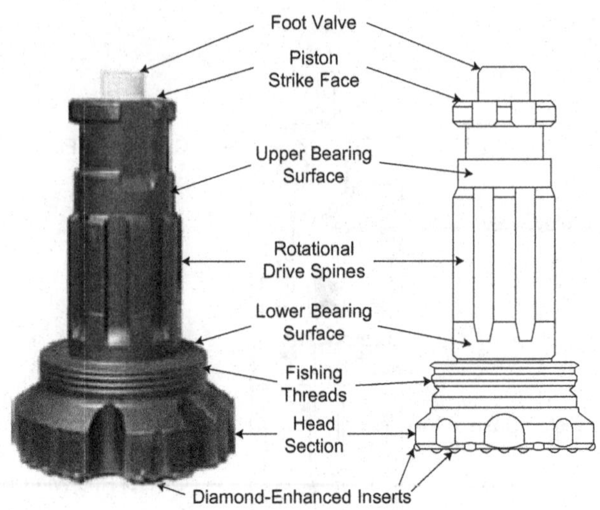

Figure 5.10 A flat-bottom bit used in gas drilling.

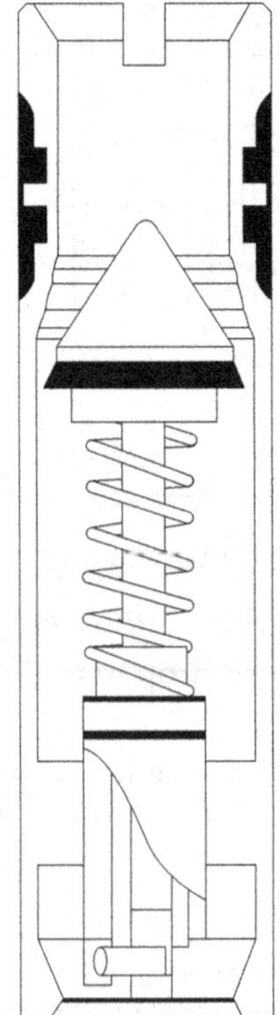

Figure 5.11 A float valve used in gas drilling.

SUMMARY

This chapter provided a brief introduction to the special equipment employed in gas drilling systems. The key pieces of equipment are gas compressors that supply powerful gas to the drilling system for hole cleaning and air hammers with flat-bottom bits.

REFERENCES

Hook, R.A., Cooper, L.W., Payne, B.R., 1977. Air, mist and foam drilling: a look at latest techniques: parts I and II. World Oil (April-May).

Lyons, W.C., Guo, B., Graham, R.L., Hawley, G.D., 2009. Air and Gas Drilling Manual, third ed. Gulf Professional Publishing.

Lyons, W.C., Guo, B., Seidel, F.A., 2001. Air and Gas Drilling Manual, second ed. McGraw-Hill.

PROBLEMS

5.1 How does a gas drilling system differ from a mud drilling system?

5.2 What are the primary and the second jets used for?

5.3 Why is the float valve necessary for gas drilling?

SUMMARY

REFERENCES

CHAPTER SIX

Gas Compressors

Contents

6.1 INTRODUCTION

Compressors are used to provide pressurized gas in air and gas drilling operations. Portable compressors were first utilized in the late 1880s in the mining industry to drill boreholes (Singer, 1958a). Deep petroleum and natural wells were drilled utilizing portable air compressors in the 1920s (Singer, 1958b). The more popular use of air as a circulating drilling fluid began in the early 1950s (Martin, 1952). By the late 1970s, it was estimated that air and gas technology was being used on about 10% of the deep wells drilled and completed (Lyons et al., 2001; GRI, 1997). This chapter addresses the technical issues involved in selecting gas compressors, including volumetric requirements, pressure requirements, and power requirements.

6.2 GAS FLOW MECHANICS

The gases used in air and gas drilling include air, natural gas, and nitrogen. Gas drilling becomes mist or unstable foam drilling when water and foaming agents are added to the stream of injection gas to increase the cuttings-carrying capacity of the gas. The liquid phase volume is less than 3% in the system. The study of the hydraulics of gas carrying solid particles (drill cuttings) is called pneumatics. The principle of pneumatics is governed by the first law of thermodynamics (conservation of energy), which is stated as

$$\Delta P = P_1 - P_2 = \frac{g}{g_c}\rho\Delta z + \frac{\rho}{2g_c}\Delta u^2 + \frac{f\rho v^2 L}{2g_c D_H} \tag{6.1}$$

where

ΔP = pressure drop, lbf/ft^2 or N/m^2
P_1 = pressure at point 1, lbf/ft^2 or N/m^2
P_2 = pressure at point 2, lbf/ft^2 or N/m^2
g = gravitational acceleration, 32.17 ft/s^2 or 9.81m/s^2
g_c = unit conversion factor, 32.17 lbm-ft/lbf-s^2 or 1 kg-m/N-s^2
ρ = fluid density, lbm/ft^3 or kg/m^3
ΔZ = elevation increase, ft or m
v = gas velocity, ft/s or m/s
f = friction factor
L = length, ft or m
D_H = hydraulic diameter, ft or m

The first, second, and third terms on the right side of Eq. (6.1) represent pressure drops due to changes in elevation, kinetic energy, and friction, respectively. The second term is usually negligible for gas flow in the pipe but is very significant for gas flow through the bit and choke.

Equation (6.1) is applicable to any flow configurations, including downward, upward, horizontal, and deviated flows. In the following sections, conventional circulation is discussed, where gas flows downward inside the drill string and upward in the annulus.

6.2.1 Gas Flow in Vertical Holes

Most wells drilled with gas in the oil and gas industry are vertical wells. The percentage of deviated and horizontal wells has been increasing rapidly since the 1980s. However, because of the limitations of directional drilling tools such as measurement while drilling (MWD) in compressible fluid drilling, gas drilling technologies are mostly used for vertical wells in the petroleum industry.

Consider an infinitesimal element of a conduit at depth H, neglecting the kinetic energy term. Eq. (6.1) gives

$$dP = \gamma_m\left(1 \pm \frac{fv^2}{2g_c D_H}\right)dH \tag{6.2}$$

where the positive sign (+) represents upward flow and the negative sign (–) represents downward flow. The γ_m, the specific weight of the mixture at depth H, is expressed as

$$\gamma_m = \frac{\dot{W}_s + \dot{W}_g + \dot{W}_l}{Q_s + Q_g + Q_l} \tag{6.3}$$

where

$\dot{W}_s$ = weight flow rate of solid, lb/sec or N/sec
$\dot{W}_g$ = weight flow rate of gas, lb/sec or N/sec
$\dot{W}_l$ = weight flow rate of liquid, lb/sec or N/sec
Q_s = volumetric flow rate of solid, ft^3/sec or m^3/sec
Q_g = volumetric flow rate of gas, ft^3/sec or m^3/sec
Q_l = volumetric flow rate of liquid, ft^3/sec or m^3/sec

The sum of the volumetric flow rates of solid and liquid is usually less than 5% of the total volumetric flow rate in air, gas, mist, and unstable foam. Equation (6.3) can be simplified to

$$\gamma_m = \frac{\dot{W}_s + \dot{W}_g + \dot{W}_l}{Q_g} \tag{6.4}$$

The volumetric flow rate of gas is related to the weight flow rate of gas through the ideal gas law:

$$Q_g = \frac{53.3\dot{W}_g(T_s + GH)}{S_g P} \tag{6.5}$$

where

T_s = ambient temperature, °R or °K
G = geothermal gradient, °F/ft or °C/m
S_g = gas specific gravity, air = 1.0

The weight rate of solids depends on bit diameter (d_b), rate of penetration (R_p), and specific gravity of solids (S_s):

$$\dot{W}_s = 62.4\frac{\pi}{4}\left(\frac{d_b}{12}\right)^2 S_s\left(\frac{R_p}{3{,}600}\right) = 9.45\times10^{-5}d_b^2 S_s R_P \tag{6.6}$$

where

d_b = bit diameter, in or m
S_s = specific gravity of solid related to freshwater
R_p = rate of penetration, ft/hr or m/hr

The weight rate of liquid depends on the misting water rate and the formation fluid influx rate:

$$\dot{W}_l = 62.4S_x\left(\frac{5.616Q_x}{3{,}600}\right) + 62.4S_l\left(\frac{5.615Q_f}{3{,}600}\right)$$
$$\dot{W}_l = 9.74\times10^{-2}(S_xQ_x + S_lQ_f) \tag{6.7}$$

where

S_x = specific gravity of misting liquid related to freshwater
Q_x = misting liquid flow rate, bbl/hr or m^3/hr
S_l = specific gravity of formation fluid related to freshwater
Q_f = formation fluid influx flow rate, bbl/hr or m^3/hr

The weight rate of gas depends on the volumetric gas flow rate Q_g (scf/min) and the specific gravity of gas S_g:

$$\dot{W}_g = 0.0765S_g\left(\frac{Q_g}{60}\right) = 1.275\times10^{-3}S_gQ_g \tag{6.8}$$

Substituting Eqs. (6.5) through (6.8) into Eq. (6.4) gives

$$\gamma_m = \frac{S_gP}{53.3(T_s+GH)}\left[1+\frac{0.074d_bS_sR_p+76.3(S_xQ_x+S_lQ_f)}{S_gQ_{go}}\right] \tag{6.9}$$

The flow velocity of gas can be formulated based on the volumetric gas flow rate, the flow path cross-sectional area, and the ideal gas law:

$$v_g = 9.77\frac{Q_g(T_s+GH)}{AP} \tag{6.10}$$

Substituting Eqs. (6.9) and (6.10) into Eq. (6.2) yields

$$dP = \frac{S_gP}{53.3(T_s+GH)}\left[1+\frac{0.074d_b^2S_sR_p+76.3(S_xQ_x+S_lQ_f)}{S_gQ_g}\right]$$
$$\times\left\{1\pm\frac{f\left[9.77\dfrac{Q_{go}(T_s+GH)}{AP}\right]^2}{2g_cD_H}\right\}dH \tag{6.11}$$

When D_H (feet) is replaced by d_H (inches), this equation can be simplified to

$$dP = \left[\frac{aP}{T_s + GH} + \frac{ab(T_s + GH)}{P}\right] dH \tag{6.12}$$

where

$$a = \frac{S_g Q_g + 0.074 d_b^2 S_s R_p + 76.3(S_x Q_x + S_l Q_f)}{53.3 Q_g} \tag{6.13}$$

and

$$b = \pm 572.7 \frac{f Q_g^2}{g A^2 d_H} \tag{6.14}$$

The solution to Eq. (6.12) is

$$(a - G)\frac{P^2}{(T_s + GH)^2} + ab = [C(T_s + GH)]^{\frac{2(a-G)}{G}} \tag{6.15}$$

where C is an integration constant. It can be determined, using the boundary condition at the top of the hole section (i.e., at depth $H = 0$), that

$$P = P_s \tag{6.16}$$

and

$$T = T_s \tag{6.17}$$

When these conditions are applied, Eq. (6.15) results in

$$P = \sqrt{\left(P_s^2 + \frac{ab}{a-G} T_s^2\right)\left(\frac{T_s + GH}{T_s}\right)^{\frac{2a}{G}} - \frac{ab}{a-G}(T_s + GH)^2} \tag{6.18}$$

which is a general equation for pressure at the point of interest in the hole at depth H.

Angel (1957) applied Weymouth's (1912) friction factor, which was derived before the friction factor was fully understood, to vertical flow when he developed his annular pressure equation. The friction factor was found to be a function of pipe wall roughness in the 1930s and 1940s when Nikuradse's (1933) correlation and Moody's (1944) chart were developed. Unfortunately, when Angel developed his model in 1957, he did not use the friction factor provided by either source. It is well known today that Moody's friction factor chart should be used whenever possible. However, it is difficult to use the chart directly when a large amount of computation is involved.

In this situation, a correlation is more convenient to use than the chart because it is easy to program in a computer. It is generally believed that in gas drilling the fluid flow in the annulus falls into the complete turbulent flow regime. In this flow regime, the friction factor is a strong function of the relative roughness and a weak function of the Reynolds number. Nikuradse's friction factor correlation is still the best available for fully developed turbulent flow in rough pipe:

$$f = \left[\frac{1}{1.74 - 2\log\left(\frac{2e}{d_H}\right)} \right]^2 \tag{6.19}$$

where

e = the absolute wall roughness in in or m

Equation (6.19) is valid for large values of the Reynolds number where the effect of relative roughness is dominant, which is consistent with Moody's chart. The major difficulty in determining the friction factor in gas drilling is estimating the absolute roughness of the wall of open holes. Although examining formation core samples indicates that for most formation rocks the absolute roughness of the drilled rock surface looks similar to that of a coarse concrete road, which has an absolute roughness of 0.06 to 0.12 in, the absolute roughness of the open holes does not necessarily fall into this range. Examining the cuttings should help locate the absolute roughness of the drilled hole.

Mason and Woolley (1981) reported that cuttings recovered at the surface are generally fine- or dust-sized particles. But there is a possibility that big cuttings that are not removed from the vicinity of the bit by the circulating air are reground by the action of the bit teeth. Large chips usually are recovered while drilling shallow holes, in deeper holes when misting and foaming, and from uphole cavings. Caliper logs indicate that wellbores normally are enlarged 0 to 15% due to fluid washout. Washout is even more notable in air drilling. Assuming a 7.5% wellbore enlargement, a 7⅞-in drill bit should generate a borehole with a wall roughness of about 0.3 in.

The average roughness of an annulus can be estimated using the following formula:

$$\bar{e} = \frac{\left(\frac{e_p d_i + e_h d_b}{d_i + d_b}\right)(H - H_c) + e_p H_c}{H} \tag{6.20}$$

where

d_i = ID of annulus (OD of drill pipe), in or m
e_p = roughness of commercial steel drill pipe and casing, in or m (a minimum value of 0.0018 in should be used if the tool joints of the drill pipe are not considered)
e_h = roughness of borehole, in or m
H_c = depth of casing shoe, ft or m

6.2.2 Gas Flow in Deviated Holes

Deviated holes represent hole sections below the kickoff points (KOPs) of directional or horizontal wells. The angle-building and angle-dropping sections can be divided into a series of slant hole segments with different inclination angles. For a small length of hole, Eq. (6.1) degenerates to

$$dP = \gamma_m dH \pm \frac{\gamma_m f v^2}{2gD_H} dS \tag{6.21}$$

where

S = segment length, ft or m

The depth and length incrementals are related through the inclination angle by

$$dH = \cos(I_s) dS \tag{6.22}$$

where

I_s = inclination angle of slant hole section, radians

Substituting Eq. (6.22) into Eq. (6.21) gives

$$dP = \gamma_m \left(\cos(I) \pm \frac{f v^2}{2gD_H} \right) dS \tag{6.23}$$

Substituting Eqs. (6.9) and (6.10) into Eq. (6.23) yields

$$dP = \frac{S_g P}{53.3[T_s + G\cos(I_s)S]}\left[1 + \frac{0.074 d_b^{\,2} S_s R_p + 76.3(S_x Q_x + S_l Q_f)}{S_g Q_{go}}\right]$$

$$\times \left\{ \cos(I_s) \pm \frac{f\left[9.77\dfrac{Q_{go}[T_s + G\cos(I_s)S]}{AP}\right]^2}{2gD_H} \right\} dS \tag{6.24}$$

which can be simplified as

$$dP = \left\{ \frac{a\cos(I_s)P}{T_s + G\cos(I_s)S} + \frac{ab[T_s + G\cos(I_s)S]}{P} \right\} dS \tag{6.25}$$

Using the boundary condition at the top of the segment—that is, $P = P_s$ at $S = 0$—the solution to Eq. (6.25) for nonhorizontal flow is

$$P = \sqrt{\left(P_s^2 + \frac{ab}{(a-G)\cos(I_s)}T_s^2\right)\left(\frac{T_s + G\cos(I_s)S}{T_s}\right)^{\frac{2a}{G}}}$$
$$\sqrt{-\frac{ab}{(a-G)\cos(I_s)}[T_s + G\cos(I_s)S]^2} \tag{6.26}$$

It is obvious that this equation degenerates to Eq. (6.18) for vertical flow when the inclination angle is zero. But it is not valid for horizontal flow where the inclination angle is 90 degrees. For horizontal flow, Eq. (6.25) becomes

$$dP = \frac{abT_s}{P}dS \tag{6.27}$$

Using the boundary condition at the exit point—that is, $P = P_s$ at $S = 0$—the solution of Eq. (6.27) for horizontal flow is

$$P^2 = P_s^2 + 2abT_s S \tag{6.28}$$

If the angle-building section has a constant radius of curvature R, there is no need to divide the curve section into a series of slant hole segments with different inclination angles. Guo and colleagues (1994) presented the following solution to gas pressure at the bottom of an arc section below the KOP:

$$P = \sqrt{P_K{}^2 + 2abRT_{av}I\exp\left[\frac{2aR\sin(I)}{T_{av}}\right]} \tag{6.29}$$

where

P_K = pressure at the KOP, lbf/ft^2 or N/m^2
T_{av} = average temperature in the arc section, R° or K°
I = inclination angle at the bottom of the arc section, radians

6.2.3 Gas Flow through the Bit

Nozzles are normally not installed on drill bits in gas drilling. Part of the reason is to reduce the problems of hole washout and "frozen" bits (see Chapter 7). Still, a significant amount of pressure drop can occur at the bit. It is preferable to have a subsonic flow of gas through the drill bits for pressure communication between the annular space and the standpipe so the pressure changes due to solid accumulation in the annulus can be identified by reading the standpipe pressure gauge (see Chapter 7).

Assuming an isentropic process, based on the second term on the right side of Eq. (6.1), the gas flow through the bit orifice obeys the following relation under subsonic flow conditions:

$$Q_g = 6.02CA_oP_{up}\sqrt{\frac{k}{(k-1)S_gT_{up}}\left[\left(\frac{P_{dn}}{P_{up}}\right)^{\frac{2}{k}} - \left(\frac{P_{dn}}{P_{up}}\right)^{\frac{k+1}{k}}\right]} \tag{6.30}$$

where

C = flow coefficient, approximately 0.6 for bit orifices
A_o = total bit orifice area, in^2 or m^2
P_{up} = upstream pressure, lb/ft^2 or N/m^2
P_{dn} = downstream pressure, lb/ft^2 or N/m^2
T_{up} = upstream temperature, R or K
k = specific heat ratio of gas, 1.4 for air and 1.28 for natural gas

Because the gas specific heat ratio is not an integer, a numerical method must be used to solve Eq. (6.30) for upstream or downstream pressures.

6.3 GAS INJECTION RATE REQUIREMENTS

Compressors should be carefully selected for gas drilling to deliver adequate gas flow rates that meet the requirements of the well being drilled. It is vitally important to maintain the volumetric flow rate of gas in an optimum range in gas drilling operations. A low gas injection rate often results in insufficient cuttings-carrying capacity and pipe sticking, while a high gas injection rate means renting large, expensive compressors and experiencing excessive wellbore washout problems. This section discusses the criteria, theory, and procedure for determining the minimum volume requirements in air, gas, mist, and unstable foam drilling. Only the direct circulation method is considered.

6.3.1 Criteria for Hole Cleaning

There are several criteria and methods for determining minimum gas volume requirements that have been used in the gas drilling industry. They fall into two categories: the minimum velocity criterion and the minimum kinetic energy criterion.

The Minimum Velocity Criterion

The minimum velocity criterion considers the interactions among particles, fluids, and the boundary of the flow domain (borehole wall). It uses the concept of terminal velocity to determine the minimum required gas velocity at the deepest large annulus.

Consider a situation where a solid particle is released in a steady, still fluid of lower density. The particle first accelerates under the action of gravity and then decelerates due to the increasing drag force on the particle from the fluid. We can prove mathematically that it will take an infinite time for the particle to reach a constant velocity. However, in reality, after a short period of time, the variation in particle velocity is not practically detectable, and the velocity of the particle reaches a "constant" velocity, also known as terminal velocity, free-settling velocity, and slip velocity.

The terminal velocity of a particle is influenced by many factors, including the size, shape, and density of the particle; the density and viscosity of the fluid; the flow regime; the particle–particle interaction; and the particle–wall interaction. Many mathematical models have been proposed to account for the effects of these factors. Assuming spherical particles, Gray (1958) presents the following equation to determine terminal settling velocity:

$$v_{sl} = \sqrt{\frac{4gD_s(\rho_s - \rho_g)}{3\rho_g C_D} \frac{\psi}{1 + D_s/D_H}} \tag{6.31}$$

where

v_{sl} = terminal settling velocity, ft/s or m/s
D_s = equivalent solid particle diameter, ft or m
ρ_s = density of solid particle, lbm/ft^3 or kg/m^3
ρ_g = density of gas, lbm/ft^3 or kg/m^3
C_D = drag coefficient accounting for the effect of particle shape: 1.40 for flat particles (shale and limestone) and 0.85 for angular to subrounded particles (sandstone)
ψ = particle sphericity factor, dimensionless
D_H = hydraulic diameter of flow path, ft or m

If no other data are available, the maximum particle size can be estimated based on the maximum penetration depth per bit revolution:

$$D_s \approx \frac{R_p}{60N} \tag{6.32}$$

where

R_p = rate of penetration, ft/hr or m/hr
N = rotary speed of bit, rpm

The minimum required gas velocity to transport the solid particles upward can be formulated as follows:

$$v_g = v_{sl} + v_{tr} \tag{6.33}$$

where

v_g = gas velocity, ft/s or m/s
v_{tr} = required particle transport velocity, ft/s or m/s

The required particle transport velocity depends on how fast the particles are generated by the drill bit and the amount of moving particles allowed in the borehole during drilling. The volumetric solid flow rate at which the particles are generated by the bit is expressed as

$$Q_s = \frac{\pi}{4}\left(\frac{d_b}{12}\right)^2 \left(\frac{R_p}{3{,}600}\right) = 1.52 \times 10^{-6} d_b^2 R_p \tag{6.34}$$

where

Q_s = volumetric flow rate of solid particles, ft^3/sec or m^3/s
d_b = bit diameter, in or m

The unit conversion factor 12 becomes 1 in SI units. The volumetric flow rate at which the solid particles are transported in the flow path is expressed as

$$Q_{tr} = v_{tr} C_p \left(\frac{A}{144}\right) \tag{6.35}$$

where

A = cross-sectional area of annular space, in^2 or m^2
Q_{tr} = volumetric flow rate of transported particles, ft^3/sec or m^3/sec
C_p = particle concentration in the flow path, volume fraction

Based on the material balance for solid particles, the volumetric flow rate of particle transport must be equal to the volumetric flow rate of particles generated by the drill bit:

$$Q_{tr} = Q_s \tag{6.36}$$

Substituting Eqs. (6.34) and (6.35) into Eq. (6.36) gives

$$v_{tr} = \frac{\pi d_b^2}{4 C_p A}\left(\frac{R_p}{3{,}600}\right) \tag{6.37}$$

Bradshaw (1964) concludes that at solid concentrations in excess of volume fraction 0.04, the tendency for solids in air to slug and interfere materially with each other becomes critical. This indicates that the critical particle concentration may be assumed to be C_p = 0.04.

Once the minimum required gas velocity is determined from Eq. (6.33), the required minimum in situ air flow rate at the collar shoulder can be estimated by

$$Q_g = 60\left(\frac{A}{144}\right) v_g \tag{6.38}$$

The required minimum in situ air flow rate is converted to the gas flow rate at the standard condition (14.7 psia, 60F) using the ideal gas law:

$$Q_{go} = \frac{T_o P}{P_o T} Q_g \tag{6.39}$$

where

Q_{go} = volumetric flow rate of gas in the standard condition, scf/min or scm/min

Since this equation still involves in situ pressure P, it has to be combined with Eq. (6.18) to solve the minimum required gas flow rate Q_{go}.

Illustrative Example 6.1

A well is cased from the surface to 7,000 ft with API 8⅝-in-diameter, 28-lb/ft nominal casing. It is to be drilled ahead to 10,000 ft with a 7⅞-in-diameter rotary drill bit, using air as a circulating fluid at an ROP of 60 ft/hr and a rotary speed of 50 rpm. The drill string is made up of 500 ft of 6¾-in OD by 213/16-in ID drill collars and 9,500 ft of API 4½-in-diameter, 16.60-lb/ft nominal EU-S135, NC 50 drill pipe. The bottomhole temperature is expected to be 160°F. We assume in this example that the annular pressure at the collar shoulder is 85 psia. Calculate the minimum required gas injection rate when the bit reaches the total depth (TD), using the minimum velocity criterion.

Solution

The maximum particle size can be estimated based on the maximum penetration depth per bit revolution:

$$D_s \approx \frac{(60)}{(60)(50)} = 0.02\,\text{ft}\ [0.006\,\text{m}]$$

Assuming a spherical sandstone particle has a specific gravity of 2.6, the terminal settling velocity can be estimated as

$$v_{sl} = \sqrt{\frac{(4)(32.2)(0.02)[(62.4)(2.6) - 0.37]}{(3)(0.37)(0.85)} \frac{1.0}{1 + \frac{(0.02)(12)}{(7.875 - 4.5)}}}$$
$$= 20.96\,\text{ft/s}[6.39\,\text{m/s}]$$

The required cuttings transport velocity can be estimated as

$$v_{tr} = \frac{\pi(7.875)^2}{4(0.04)\left[\frac{\pi}{4}(7.875^2 - 4.5^2)\right]}\left(\frac{60}{3{,}600}\right)$$
$$= 0.62\,\text{ft/s}[0.19\,\text{m/s}]$$

The gas velocity required to transport the solid particles can be calculated as

$$v_g = 20.96 + 0.62 = 21.58\,\text{ft/s}\ [6.58\,\text{m/s}]$$

(Continued)

Illustrative Example 6.1 *(Continued)*

The required minimum volumetric air flow rate at the collar shoulder is estimated to be

$$Q_g = \frac{\pi}{4}\left(\frac{(7.875)^2 - (4.5)^2}{144}\right)(21.58)(60) = 295\,\text{ft}^3/\text{min} = 8.35\text{m}^3/\text{min}$$

The required minimum volumetric air flow rate at the collar shoulder is converted to standard condition as

$$Q_{go} = \frac{(520)(85)(144)(295)}{(14.7)(144)(620)} = 1{,}431\,\text{scf/min} = 40.52\,\text{scm/min}$$

The Minimum Kinetic Energy Criterion

The minimum kinetic energy criterion was established in the 1950s. The mixture of gas and solid is treated as one homogeneous phase with mixture density and velocity—that is, interactions between particles and fluids are not considered. Several models have been presented, including those by Martin (1952, 1953), Scott (1957), Angel (1957), and McCray and Cole (1959). Although McCray and Cole's model permits a constant-percentage slip velocity of solid particles, it uses the same particle lift criterion as Angel's model.

The criterion for the minimum volume requirement is based on the experience gained from quarry drilling with air. The minimum annular velocity to effectively remove solid particles from the borehole is usually assumed to be 3,000 ft/min, or 50 ft/sec (ft/s), under atmospheric conditions (close to the standard condition of 14.7 psia at 60°F). This velocity is believed to be high enough to remove dustlike particles in air drilling. Although big cuttings that are not removed from the vicinity of the bit by the circulating air are reground by the bit teeth, it would be uneconomical to lift large cuttings without first trying to control their initial size at the bit.

The carrying power of air can be evaluated based on its kinetic energy E_o per unit volume of air:

$$E_{go} = \frac{1}{2}\frac{\gamma_{go}}{g}v_{go}^2 \tag{6.40}$$

where

γ_{go} = specific weight of standard air (0.0765 lb/scf or 12 N/scm)
v_{go} = minimum required velocity of air under standard conditions (50 ft/s or 15.24 m/s)

The kinetic energy of 1 standard cubic foot of air moving at a velocity of 50 ft/s is

$$E_{go} = \frac{1}{2}\left(\frac{0.0765}{32.2}\right)(50)^2 \approx 3 \text{ ft-lb/ft}^3[143\,\text{J/m}^3]$$

If the carrying capacity of gas at the point of interest in the borehole is equivalent to the carrying power of the velocity of standard air, the following relationship must hold:

$$\frac{1}{2}\frac{\gamma_g}{g}v_g^2 = \frac{1}{2}\frac{\gamma_{go}}{g}v_{go}^2 \tag{6.41}$$

where

γ_g = specific weight of in situ gas, lb/ft^3 or N/m^3
v_g = velocity of in situ gas, ft/s or m/s

The specific weight of gas in Eq. (6.41) can be expressed as a function of in situ pressure and temperature based on the ideal gas law:

$$\gamma_g = \frac{S_g P}{53.3T} \tag{6.42}$$

The volumetric in situ flow rate of gas can be determined based on Eq. (6.39):

$$Q_g = \frac{P_o T}{T_o P} Q_{go} \tag{6.43}$$

Dividing Eq. (6.43) by the cross-sectional area of the flow path yields an expression for in situ gas velocity in U.S. engineering units as

$$v_g = \frac{(14.7)(144)(144)TQ_{go}}{(60)(520)PA} \tag{6.44}$$

or

$$v_g = 9.77\frac{TQ_{go}}{PA} \tag{6.45}$$

Substituting Eqs. (6.42) and (6.45) into Eq. (6.41) and rearranging the latter yields

$$Q_{go} = \frac{v_{go}A}{4.84}\sqrt{\frac{P}{23.41S_gT}} \quad (6.46)$$

Since this equation still involves in situ pressure P, it must be combined with Eq. (6.18) to solve the minimum required gas flow rate Q_{go}. This approach can be used to generate engineering charts for various well conditions. Some of the charts are presented in Appendix B.

The results of Illustrative Examples 6.1 and 6.2 indicate that the calculated minimum required air injection rate given by the minimum velocity criterion is slightly lower than that given by the minimum kinetic energy criterion, even though a very large particle size (nearly ¼ inch) is used. Although the minimum velocity criterion appears to be more general, difficulties associated with the rough estimation of the unknown

Illustrative Example 6.2

Solve the problem in Illustrative Example 6.1 using the minimum kinetic energy criterion.

Solution

The in situ gas specific weight can be calculated as

$$\gamma_g = \frac{(1)(85)(144)}{53.3(460+160)} = 0.37\ \text{lb/ft}^3$$

The minimum air velocity value can be calculated as

$$\frac{1}{2}\left(\frac{0.37}{32.2}\right)v_g^2 = \frac{1}{2}\left(\frac{0.0765}{32.2}\right)(50)^2$$

which gives in situ gas velocity of $v_g = 22.6$ ft/s [6.58 m/s]. The required minimum in situ air flow rate is estimated to be

$$Q_g = \frac{\pi}{4}\left(\frac{(7.875)^2-(4.5)^2}{144}\right)(22.6)(60) = 309\ \text{ft}^3/\text{min}\ [8.76\text{m}^3/\text{min}]$$

The required minimum in situ air flow rate is converted to standard conditions using the ideal gas law as

$$Q_{go} = \frac{(520)(85)(309)}{(14.7)(620)} = 1{,}499\ \text{scf/min}\ [42.48\ \text{scm/min}]$$

parameters (e.g., the shape and size of the particles) have hindered its practical application.

Schoeppel and Spare (1967) reported that the gas flow rate values obtained from the minimum kinetic energy criterion were at least 25% below the actual field's needs. This motivated numerous investigators to develop more accurate models to determine the minimum required gas injection rate for gas drilling. These models include those presented by Capes and Nakamura (1973), Sharma and Crowe (1977), Ikoku and colleagues (1980), Machado and Ikoku (1982), Mitchell (1983), Puon and Ameri (1984), Sharma and Chowdry (1984), Wolcott and Sharma (1986), Adewumi and Tian (1989), Tian and Adewumi (1991), and Supon and Adewumi (1991).

Guo and colleagues (1994) performed a comparison of results from model calculations and field experience. The comparison showed that among these models, only the result given by Angel's model obtained from the minimum kinetic energy criterion has a trend consistent with field experience, although Angel's model provides lower estimates for the minimum volumetric gas requirements. Guo and colleagues found that Angel's charts give values lower than field requirements because Weymouth's friction derived for flow in smooth pipes was used in Angel's calculations for flow in rough wellbores.

Guo and colleagues incorporated Nikuradse's (1933) friction factor correlation for rough boreholes into Angel's analysis. This improvement made Angel's approach as far as field experience the best. Guo and colleagues also introduced various hole inclinations into the analysis and used the concept of the kinetic energy index as an indicator for hole cleaning. The kinetic energy index is defined as the kinetic energy of the in situ gas divided by the minimum required kinetic energy (3 ft-lb/ft^3).

Guo and Ghalambor (2002) generated correlations and engineering charts for determining the air and gas flow rates required for hole cleaning using the minimum kinetic energy criterion. Lyons and colleagues (2009) coded the minimum kinetic energy criterion in their MathCad programs.

6.3.2 Corrections for Site Pressure, Temperature, and Humidity

When a compressor is operated at surface locations above sea level, the volumetric air flow rate intake is the actual cubic feet per minute (acf/min), not the standard cubic feet per minute (scf/min). The ambient pressure and

Table 6.1 Atmospheric Pressures and Temperatures at Different Elevations

Elevation (ft)	Pressure (psia)	Temperature (°F)
0	14.696	59.00
2,000	13.662	51.87
4,000	12.685	44.74
6,000	11.769	37.60
8,000	10.911	30.47
10,000	10.108	23.36

temperature of the atmosphere decrease as elevation increases. The decrease in atmospheric pressure reduces the mass flow rate of gas at the suction end of the compressor, while the drop in temperature increases the mass flow rate of the gas.

Table 6.1 gives the average atmospheric pressure and temperature for latitudes from 30°N to 60°N. The temperature data in the table should be used with caution because onsite temperatures vary significantly with seasons.

The minimum required volumetric flow rate of site air should be determined based on Q_{go} and the site atmospheric pressure and temperature using the ideal gas law:

$$Q_a = \frac{0.0283 T_a}{p_a} Q_{go} \tag{6.47}$$

where

p_a = actual atmospheric pressure at the drilling site, psia or kPa
T_a = actual atmospheric temperature at the drilling site, °R or °K

Corrections for site humidity should also be made. The density of water vapor is less than the density of air under the same pressure and temperature. Consequently, the density of humid air is less than the density of dry air, and humid air contains less mass than dry air. More importantly, water vapor in the air is usually removed in the after-cooling system of compressors to reduce the detrimental effects of freshwater on borehole conditions. Even though its removal is incomplete, the remaining water vapor can be liquefied at the bit. When small bit orifices are used, the temperature at the bit can be lower than the dew point and even the ice point of water due to the Joule-Thomson effect (discussed in Chapter 7).

Miska (1984) presents a formula to correct air humidity. If the water-separating efficiency of compressors is considered, Miska's formula can be modified to

$$Q_h = \frac{p_a}{p_a - f_w \Phi p_w} Q_a \tag{6.48}$$

where

Q_h = volumetric flow rate of humid air, ft^3/min or m^3/min
f_w = water separation efficiency, fraction
Φ = relative humidity, fraction
p_w = water vapor saturation pressure, psia or kPa

The water vapor saturation pressure can be estimated based on temperature from the following formula (Miska, 1984):

$$p_w = 10^{6.39416 - \frac{1750.286}{217.23 + 0.555t}} \tag{6.49}$$

where

t = temperature, °F

Illustrative Example 6.3

A well is to be drilled with air at an elevation of 4,000 ft. The site air relative humidity is 0.8 at a temperature of 85°F. The compressor's dewatering efficiency is 95%. The minimum required air injection rate is estimated to be 2,485 scf/min to carry up the cuttings. Determine the minimum required compressor capacity (in situ air flow rate) for the operation.

Solution

Based on Table 6.1, the site pressure is estimated to be 12.685 psia. The dry air requirement is calculated as

$$Q_a = \frac{(0.0283)(85 + 460)}{(12.685)}(2{,}485) = 3{,}021\ \text{ft}^3/\text{min}$$

The water saturation pressure is

$$p_w = 10^{6.39416\ \frac{1750.286}{217.23 + (0.555)(85)}} = 0.5949\ \text{psia}$$

The humid air requirement is calculated to be

$$Q_h = \frac{12.685}{12.685 - (0.95)(0.8)(0.5949)}(3{,}021) = 3{,}133\ \text{ft}^3/\text{min}$$

6.4 GAS INJECTION PRESSURE REQUIREMENTS

Compressors should be selected for gas drilling to deliver adequate gas pressure at the design gas injection rate. The design gas injection rate should be determined based on the flow capacity of the candidate compressor. The design gas injection rate should never be less than the minimum gas injection rate required by hole cleaning. With the design gas injection rate, the required compressor pressure can be calculated on the basis of the extreme borehole architecture and the extreme borehole conditions.

The extreme borehole architecture is usually the well configuration at the total depth to drill with gas. Well trajectory can consist of straight, deviated, or horizontal sections. The drill string configuration includes the drill pipe, the bottomhole assembly (BHA) with or without a gas hammer, and the drill bit. The extreme borehole condition is usually the situation where formation water is encountered and unstable foam is used for water removal. This requires knowing the intensity of the formation water influx.

The procedure for determining the required gas injection pressure is as follows:

1. Predict the minimum gas injection rate based on the hole cleaning requirement.
2. Find a compressor that has a gas capacity higher than the minimum required gas injection rate with a flow rate safety factor of not less than 1.15 or a kinetic energy index of not less than 1.2.
3. Use the gas capacity of the compressor as the design gas injection rate to calculate the gas injection pressure in the extreme borehole architecture under the extreme borehole condition—specifically:
   - Calculate the flowing pressure at the kickoff point in the annulus using Eq. (6.18).
   - Calculate the flowing pressure at the end of the angle-building section in the annulus using Eq. (6.29) or repeatedly using Eq. (6.26).
   - Calculate the flowing pressure at the end of the angle-holding section in the annulus using Eq. (6.26). If the angle-holding section is horizontal, use Eq. (6.28).
   - Calculate the flowing pressure above the drill bit inside the BHA by numerically solving Eq. (6.30) for the upstream pressure.
   - Calculate the flowing pressure at the top of the drill color inside the BHA using Eq. (6.26). If the BHA is horizontal, use Eq. (6.28).

- Calculate the flowing pressure at the KOP inside the drill string using Eq. (6.29) or repeatedly using Eq. (6.26).
- Calculate the flowing pressure in the standpipe using Eq. (6.18).
- Calculate the gas injection pressure using Eq. (6.28).

Illustrative Example 6.4

A well is to be cased to 2,057 ft with a 9⁵⁄₈-in, 40-lb/ft (8.835-in ID) casing. Starting from the kickoff point at 2,084 ft, the hole will be drilled with mud using a 7⁷⁄₈-in bit to build the inclination angle at a constant build rate of 5°/100 ft until the maximum inclination angle of 74.05 degrees is reached at a depth of 3,538 ft. Then the drilling fluid will be shifted from mud to air to drill slant to the TD of 8,145 ft, while the inclination angle is maintained at 74.05 degrees. Additional data are given.

1. Assuming negligible pressure loss in the surface injection line, determine the minimum required gas injection pressure using a kinetic energy index safety factor of 1.2 for the design gas injection rate.
2. Plot profiles of the pressure, velocity, gas density, and kinetic energy index along the flow path.

Drill string data
Drill pipe OD: 4.5 in
Drill pipe ID: 3.643 in
Drill collar length: 247 ft
Drill collar OD: 6.25 in
Drill collar ID: 3 in
Material properties
Specific gravity of rock: 2.8 (water = 1)
Specific gravity of gas: 1 (air = 1)
Gas specific heat ratio (k): 1.25 (water = 1)
Specific gravity of misting fluid: 1 (water = 1)
Specific gravity of formation fluid: 1
Pipe roughness: 0.0018 in
Borehole roughness: 0.3 in
Environment
Site elevation (above mean sea level): 2,000 ft
Ambient pressure: 13.665 psia
Ambient temperature: 75°F
Relative humidity: 0.8 fraction
Geothermal gradient: 0.01 F/ft

(Continued)

Illustrative Example 6.4 *(Continued)*

Operating conditions

Surface choke/flow line pressure: 15 psia
Rate of penetration: 120 ft/hour
Rotary speed: 50 rpm
Misting rate: 10 bbl/hour
Formation fluid influx rate: 10 bbl/hour
Dewatering efficiency: 0.95 fraction 1/32nd
Bit orifices: 20 in
20-1/32nd in
20-1/32nd in

Solution

These problems can be solved with computer program *GasDrill-09.xls.*

1. A gas injection rate of 1,430 scf/min gives a kinetic energy index of 1.0 at the shoulder of the drill collar. A gas injection rate of 1,615 scf/min gives a kinetic energy index of 1.2. The minimum required injection pressure is calculated to be 136 psia.
2. Using the design gas injection rate of 1,615 scf/min, the profiles of the pressure, velocity, gas density, and kinetic energy index along the flow path are calculated and plotted in Figures 6.1, 6.2, 6.3, and 6.4, respectively.

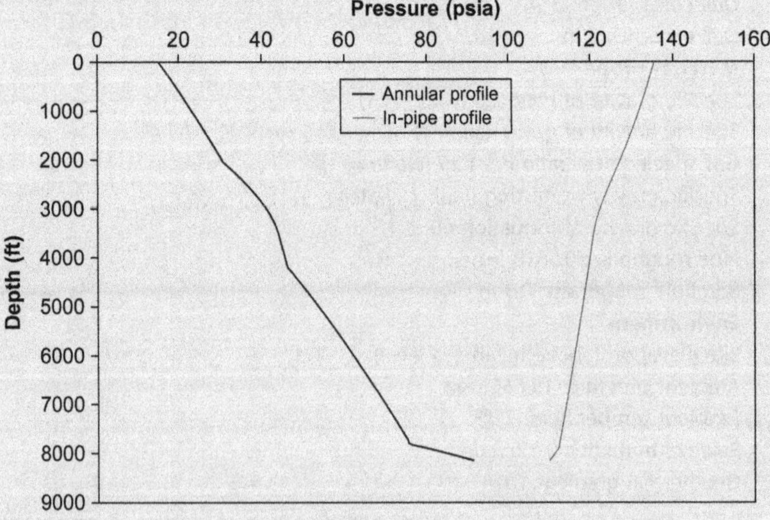

Figure 6.1 Calculated pressure profile.

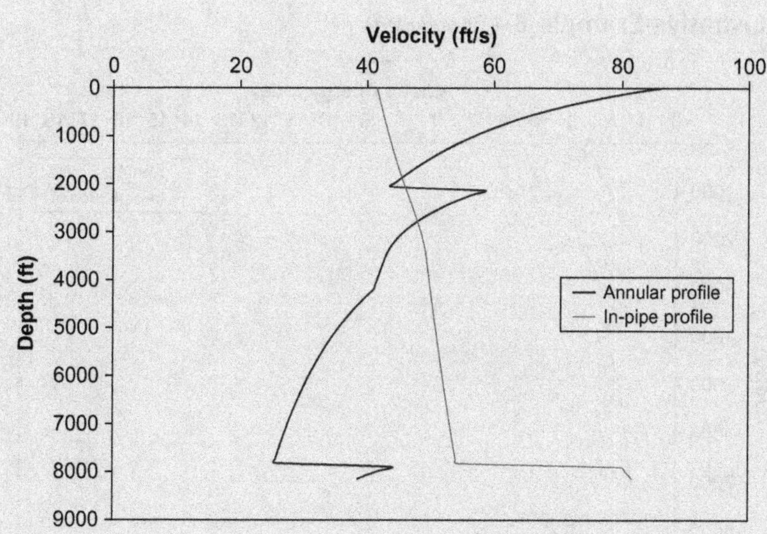

Figure 6.2 Calculated velocity profile.

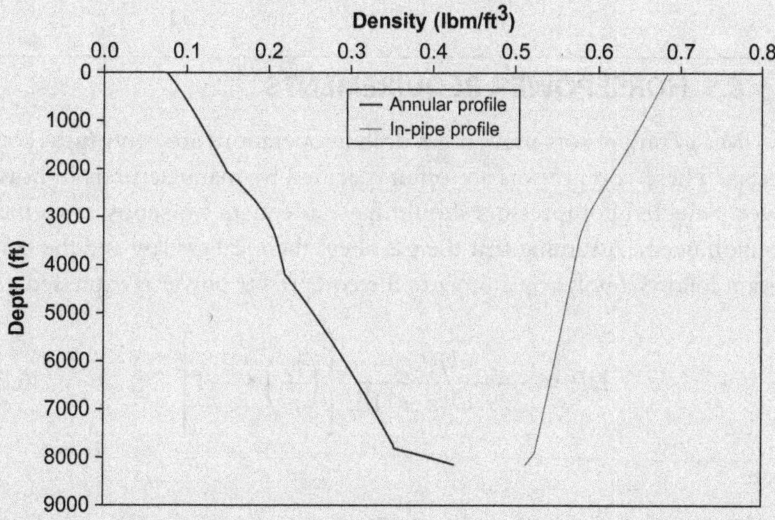

Figure 6.3 Calculated gas density profile.

(*Continued*)

Illustrative Example 6.4 *(Continued)*

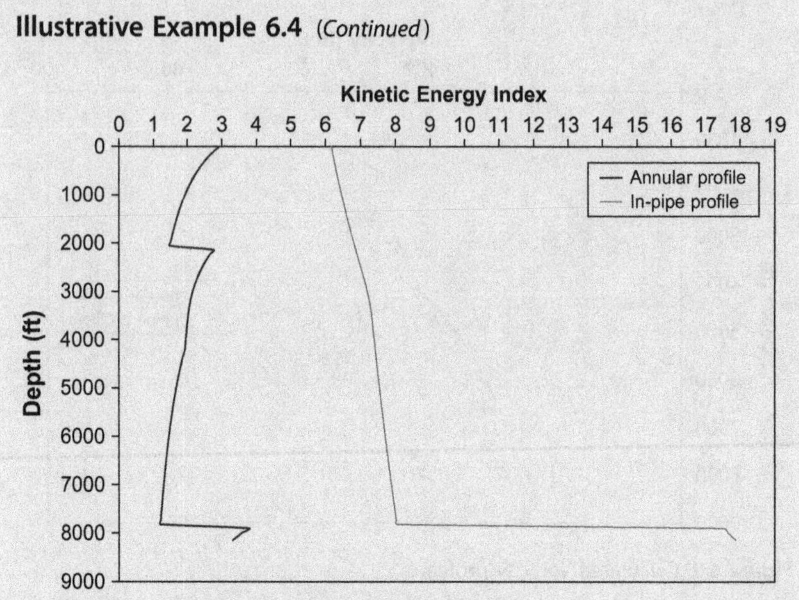

Figure 6.4 Calculated kinetic energy index profile.

6.5 HORSEPOWER REQUIREMENTS

Most compressors used in gas drilling operations are centrifugal compressors. These compressors are often specified by manufacturers in horsepower. Selected compressors should have adequate horsepower to meet operation needs. Assuming that the gas obeys the ideal gas law and the compression follows a polytropic process, the compressor power is expressed as

$$HP_c = \frac{n_s k p_i Q_{gi}}{229.17(k-1)E_p}\left[\left(\frac{p_o}{p_i}\right)^{\frac{k-1}{n_s k}} - 1\right] \tag{6.50}$$

where

HP_c = compressor power, hp or W
n_s = number of compression stages
k = specific heat ratio of gas
p_i = gas pressure at inlet condition, psia or N.m^2
Q_{gi} = gas flow rate at inlet condition, ft^3/min or m^3/min

E_p = polytropic efficiency (0.7 ~ 0.9), fraction
p_o = gas pressure at outlet condition, psia or N.m^2

It is important to know that the efficiency drops significantly with compression ratio. In field practice, the pressure ratio seldom exceeds 4. When the compression ratio is greater than 4, the compression is broken into multiple stages, with the compression ratio in each stage being less than 4. In gas drilling operations where the required gas pressure is on the order of a few hundred psia, three-stage compression is often adequate for the primary compressors. More stages of compression are used in the secondary compressors (boosters).

Illustrative Example 6.5

For the air drilling conditions specified in Illustrative Example 6.4, determine the required compressor horsepower. Assume that the efficiency of the compressor is 0.71.

Solution

The dry air requirement is calculated as

$$Q_a = \frac{0.0283(75+460)}{13.665}(1615) = 1{,}789\ \text{ft}^3/\text{min}$$

The water vapor saturation pressure can be estimated as

$$p_w = 10^{\left(6.39416 - \frac{1750.286}{217.23+0.555(75)}\right)} = 0.4291\ \text{psia}$$

The humid air requirement is calculated to be

$$Q_h = \frac{13.665}{13.665-(0.95)(0.8)(0.4291)}(1{,}789) = 1{,}833\ \text{ft}^3/\text{min}$$

The overall compression ratio is

$$\frac{p_o}{p_i} = \frac{136}{13.665} = 9.92 > 4$$

If two stages of compression are used:

$$\left(\frac{p_o}{p_i}\right)^{1/2} = \sqrt{9.92} = 3.15 < 4\ \text{OK}$$

Equation (6.50) gives a required compressor horsepower of

$$HP_c = \frac{(2)(1.25)(13.665)(1{,}833)}{229.17(1.25-1)(0.71)}\left[\left(\frac{136}{13.665}\right)^{\frac{1.25-1}{(2)(1.25)}} - 1\right] = 397\ \text{hp}$$

SUMMARY

This chapter provided a method and a procedure for selecting compressors for air and gas drilling. It involves predictions of the minimum required gas injection rate, the minimum required gas injection pressure, and the minimum required compressor power. The minimum kinetic energy criterion is recommended for hole cleaning evaluations. Local ambient pressure, temperature, and air humidity should be considered when selecting air compressors.

REFERENCES

Adewumi, M.A., Tian, S., 1989. Hydrodynamic modeling of wellbore hydraulics in air drilling. Paper SPE 19333, Proceedings of the SPE Eastern Regional Meeting, Society of Petroleum Engineers, October 24-27, Morgantown.

Angel, R.R., 1957. Volume requirements for air or gas drilling. Trans. AIME 210, 325–330.

Bradshaw, S.K., 1964. A numerical analysis of particle lift. MS thesis, University of Oklahoma.

Capes, C.E., Nakamura, K., 1973. Vertical pneumatic conveying: An experimental study with particles in the intermediate and turbulent flow regimes. Can. J. Chem. Eng. (March), 33–38.

Gas Research Institute (GRI), 1997. Underbalanced Drilling Manual. Gas Research Institute Publication, Chicago.

Gray, K.E., 1958. The cutting carrying capacity of air at pressures above atmospheric. Trans. AIME 213, 180–185.

Guo, B., Ghalambor, A., 2002. Gas Volume Requirements for Underbalanced Drilling Deviated Holes. PennWell Books.

Guo, B., Miska, S., Lee, R.L., 1994. Volume requirements for directional air drilling, Paper IADC/SPE 27510, Proceedings of the IADC/SPE Drilling Conference, February 15–18, Dallas.

Ikoku, C.U., Azar, J.J., Williams, C.R., 1980. Practical approach to volume requirements for air and gas drilling. Paper SPE 9445, Proceedings of the SPE 55th Annual Fall Technical Conference and Exhibition, September 21-24, Dallas.

Lyons, W.C., Guo, B., Seidel, F.A., 2001. Air and Gas Drilling Manual, second ed. McGraw-Hill.

Lyons, W.C., Guo, B., Graham, R.L., Hawley, G.D., 2009. Air and Gas Drilling Manual, third ed. Gulf Professional Publishing.

Machado, C.J., Ikoku, C.U., 1982. Experimental determination of solid fraction and minimum volume requirements in air and gas drilling. J. Pet. Technol. (November), 35–42.

Martin, D.J., 1952. Use of air or gas as a circulating fluid in rotary drilling—volumetric requirements. Hughes Eng. Bull. 23, 29, 35–42.

Martin, D.J., Additional calculations to determine volumetric requirements of air or gas as a circulating fluid in rotary drilling. Hughes Eng. Bull. 23-A, 22.

Mason, K.L., Woolley, S.T., 1981. How to air drill from compressor to blooey line. Pet. Eng. Int. (April), 120–136.

McCray, A.W., Cole, F.W., 1959. Oil Well Drilling Technology. University of Oklahoma Press.

Miska, S., 1984. Should we consider air humidity in air drilling operations?. Drill Bit (July), 8–9.

Mitchell, R.F., 1983. Simulation of air and mist drilling for geothermal wells. J. Pet. Technol. (November), 27–34.

Moody, L.F., 1944. Friction factor for pipe flow. Trans. ASME 66, 671–685.
Nikuradse, J., 1933. A new correlation for friction factor. Forschungshelf, 301–307.
Puon, P.S., Ameri, S., 1984. Simplified approach to air drilling operations. Paper SPE 13380, Proceedings of the SPE Eastern Regional Meeting, Society of Petroleum Engineers, October 31–November 2, Charleston.
Schoeppel, R.J., Spare, A.R., 1967. Volume requirements in air drilling. SPE Preprint 1700, Society of Petroleum Engineers.
Scott, J.O., 1957. How to figure how much air to put down the hole in air drilling. Oil & Gas J. (December 16), 104–107.
Sharma, M.P., Chowdry, D.V., 1984. A computational model for drilled cutting transport in air (or gas) drilling operations. Paper 12336, Proceedings of the Southeast Conference on Theoretical and Applied Mechanics, SPE of AIME.
Sharma, M.P., Crowe, C.T., 1977. A novel physico-computational model for quasi: one-dimensional gas-particle flows. Trans. ASME 22, 79–83.
Singer, C. *et al.*, 1958a. History of Technology, vol. 4. Oxford Press.
Singer, C. *et al.*, 1958b. History of Technology, vol. 6. Oxford Press.
Supon, S.B., Adewumi, M.A., 1991. An experimental study of the annulus pressure drop in a simulated air-drilling operation. SPE Drill. Completion J. (September), 74–80.
Tian, S., Adewumi, M.A., 1991. Development of hydrodynamic model-based air drilling design procedures. Paper SPE 23426, Proceedings of the SPE Eastern Regional Meeting, Society of Petroleum Engineers, October 22-25.
Weymouth, T.R., 1912. Problems in natural gas engineering. Trans. ASME 34, 185–189.
Wolcott, P.S., Sharma, M.P., 1986. Analysis of air drilling circulating systems with application to air volume requirement estimation. Paper SPE 15950, Proceedings of the SPE Eastern Regional Meeting, Society of Petroleum Engineers, November 12–14, Columbus.

PROBLEMS

6.1 A well is cased from the surface to 5,000 ft with API 8⅝-in diameter, 28-lb/ft nominal casing. It is to be drilled ahead to 8,000 ft with a 7⅞-in-diameter rotary drill bit using air as a circulating fluid at an ROP of 90 ft/hr and a rotary speed of 60 rpm. The drill string is made up of 400 ft of 6¾-in OD by 2$^{13}/_{16}$-in ID drill collars and 7,600 ft of API 4^{1}/-in-diameter, 16.60-lb/ft nominal EU-S135, NC 50 drill pipe. The bottomhole temperature is expected to be 140°F. It is assumed that the annular pressure at the collar shoulder is 75 psia. Calculate the minimum required gas injection rate when the bit reaches the total depth (TD) using (1) the minimum velocity criterion and (2) the minimum kinetic energy criterion.

6.2 A well is designed to drill to a TD of 9,205 ft. It is cased to 3,050 ft with a 9⅝-in, 40-lb/ft (8.835-in ID) casing. Starting from the kickoff point at 3,075 ft, the hole is to be drilled with mud using a 7⅞-in bit to build an inclination angle at a constant build rate of 4°/100 ft until the maximum inclination angle of 55 degrees is reached. Then the drilling fluid will be shifted from mud to air to drill a slant hole to the TD. The following additional data are given.

Drill string data
Drill pipe OD: 4.5 in
Drill pipe ID: 3.643 in
Drill collar length: 288 ft
Drill collar OD: 6.25 in
Drill collar ID: 2.5 in
Material properties
Specific gravity of rock: 2.75 (water = 1)
Specific gravity of gas: 1 (air = 1)
Gas specific heat ratio (k): 1.25 (water = 1)
Specific gravity of misting fluid: 1 (water = 1)
Specific gravity of formation fluid: 1 in
Pipe roughness: 0.002 in
Borehole roughness: 0.25
Environment
Site elevation (above mean sea level): 2000 ft
Ambient pressure: 13.67 psia
Ambient temperature: 65°F
Relative humidity: 0.7 fraction
Geothermal gradient: 0.01 F/ft
Minimum velocity under standard conditions: 50 ft/sec
Operating condition
Surface choke/flow line pressure: 14.7 psia
Rate of penetration: 90 ft/hour
Rotary speed: 50 rpm
Misting rate: 5 bbl/hour
Formation fluid influx rate: 15 bbl/hour
Dewatering efficiency: 0.90 fraction
Bit orifices: 20-1/32nd in
20-1/32nd in
20-1/32nd in

1. Assume that the pressure loss in the surface injection line is negligible, and predict the minimum required gas injection pressure using a gas flow rate safety factor of 1.15 for the design gas injection rate.
2. Plot the profiles for pressure, velocity, gas density, and kinetic energy index inside the drill string and in the annulus.

6.3 For the air drilling conditions specified in problem 6.2, determine the required compressor horsepower. Assume that the efficiency of the compressor is 0.75.

CHAPTER SEVEN

Gas Drilling Operations

Contents

7.1 INTRODUCTION

Gas drilling uses air, nitrogen, or natural gas to drill relatively hard and dry formation intervals. A small amount of water with surfactants is sometimes added to the gas stream to reduce drilling problems. This chapter introduces operating procedures in gas drilling to reduce operation complications and addresses measures that are usually taken to solve gas drilling problems.

7.2 GAS DRILLING PROCEDURES

Special procedures are followed in gas drilling to ensure smooth operations. These include initiating gas drilling, initiating mist drilling, making connections, and tripping pipe.

7.2.1 Initiating Gas Drilling

Compressor operators participate in gas drilling routinely. Their general procedures should be followed because they are familiar with both the local area and their compressors. To unload an air hole, the cement and cementing shoe should be drilled out with water or mud, and the cuttings should be circulated out of the hole. It may take several hours to dry a hole if the hole is not washed clean. The following procedure should be followed:

1. Go in the hole to the bottom of the casing and pump alternate slugs of gas to the pressure limit of the compressors and water (do not use mud) to bring the pressure back down again for further air injection.
2. When most of the water is out of the hole, pump 5 gallons of foaming agent into the pipe and circulate it around the hole. The detergents will bring up a large quantity of water.
3. Go to the bottom and repeat this process.
4. Light the blooey line flare or igniter.
5. Drill a little formation and then pick up the pipe past a tool joint.
6. Continue this process until the well begins producing dust.
7. If after 2 hours the well still has not dusted, then
   a. Add 5 gallons of 50% foamer/water mixture and start again at step 4, or
   b. Close the rams and pressure up the hole with air, and then open the rams and let the compressed air blow out of the hole. Then start again at step 4.
   c. Close the radiator shutters on the compressors, or place a piece of cardboard on the compressor aftercooler to increase the heat of the air going into the hole.

7.2.2 Initiating Mist Drilling

Misting gas can help hole cleaning, depress downhole fires or explosions, lower borehole friction, and reduce drill string vibrations. The following procedure is used to initiate mist drilling:

1. Unload the mud out of the hole as with air drilling but without attempting to dry the hole.
2. Turn on the mist pump and begin gas circulation. About 1 foot of formation should be drilled, and the string should be picked up past a tool joint. Continue this procedure for at least 5 feet. Go to regular mist drilling if there is no drag on the pickup after 5 feet.

3. As soon as there are steady returns, the mist quantities can be adjusted. Wait at least 30 minutes before making additional changes after any adjustment to the mist volume or mixture.
4. If too many changes are made too quickly, it is difficult to determine the impact of the adjustment.

7.2.3 Making Connections

The procedure for making connections is as follows:

1. The well should be circulated until the returns are free of cuttings or at least minimized.
2. While circulating the hole clean, reciprocate the pipe slowly to wipe out any potential mud rings that have begun to form.
3. Gas should be diverted down the blooey line, and the flow rate from the compressor should be reduced.
4. Pull up the kelly and set the slips. Open the bleed-off line and allow the gas to bleed off.
5. Slowly break the kelly loose and allow any air to be vented from the drill string.
6. Make up the next joint of drill pipe as normal.
7. Begin lowering the pipe back to the bottom and return the flow stream down the drill string. When returns are seen at the blooey line, drilling can be resumed.

7.2.4 Tripping Pipe

The following procedure should be used in pipe tripping:

1. Tripping can be considered as a series of connections. However, a major change occurs when the BHA arrives at the surface. If the drill collar/stabilizer/large OD tools can pass through the pack-off, then the trip can continue as before.
2. If they are too large to pass through the pack-off, then the diverter pack-off will have to be removed.
3. By the time the BHA is tripped to the surface, the annulus has most likely blown down. If this is the case, the trip can continue because there is minimal danger from a kick.
4. If the well is producing gas or liquids that can be successfully diverted down the flow line or blooey line by the vacuum created with the compressor flow, then the trip can be continued with precautions taken on each connection.

5. If the well cannot be controlled, then the annular preventer will need to be closed, and a liquid pill must be pumped into the annulus to create a cushion and stop the annular flow. It is possible that the well will have to be killed.
6. Once the bit has cleared the blind ram cavity, the blind ram should be closed. The pack-off can be removed after any trapped pressure is bled off.

7.3 PROBLEMS AND SOLUTIONS

Problems frequently encountered in gas drilling include borehole instability, "mud rings," water loading, bit balling, crooked holes, corrosion, and downhole fires and explosions. Special measures are taken to solve these problems.

7.3.1 Borehole Instability

Owing to the low density of gas, wellbore pressures are much lower in gas drilling than in conventional drilling operations. Lack of support of fluids to the borehole frequently causes wellbore instability, often resulting in boreholes caving in, collapsing, or getting narrower. These problems can lead to pipe sticking and failure of the drilling operations. This type of borehole stability problem may occur when rock formations are very soft or weak or if they contain significant amounts of water-sensitive clays. The borehole stability problem can be eased by mixing the gas with some salt/surfactant solutions. The solution can stabilize the borehole in two ways: by increasing the fluid pressure to support the hole wall and by inhibiting the clays from swelling.

Among many chemicals, KCl is widely used for controlling clay swelling. Anionic surfactants or household detergents are often used as foaming agents to form foam in the annulus so the inhibiting solution can be brought up to the intervals where water-sensitive clays exist. Inhibited mist (3–4% KCl and soap) can eliminate the hydration of clays. Adding polymer also helps. To prevent air slugging, use the following formula per barrel of water:

- 1/8 lb polyanionic cellulose polymer (PAC or CMC)
- 1/8 lb xanthan gum polymer (XC or CMC)
- 1% foaming agent by weight water

However, in many cases the formations are too weak or soft to be drilled with gas even if significant amounts of inhibitors are used. In these situations, the open hole needs to be cased before drilling ahead with gas, or the gas drilling needs to be converted to foam drilling operations to solve the borehole stability problem. Chemically induced borehole stability problems, such as clay swelling, get worse as time goes on. It is a good practice to drill the sensitive intervals fast and then set the casing to isolate them.

7.3.2 Mud Rings

Mud rings are a result of inadequate hole cleaning. As shown in Figure 7.1, when drill cuttings move to the shoulder of drill collars, a floating bed of cuttings may form due to the low velocity of in situ gas in the pipe–open hole annulus. Although the gas density is relatively high at this point compared to the upper hole section, the kinetic energy of gas at this point is the lowest. If the gas stream is not powerful enough to keep the cuttings floating, significant amounts of cuttings may accumulate in this area. Figure 7.2 illustrates this situation. When the formation is damp from water or oil, the cuttings form a "mud" that is deposited against the side of the hole. This tends to form rings of mud that, as they grow larger, restrict airflow and cause the pressure to increase. Mud rings cause high friction, which can result in downhole burn-offs (fires) and stuck pipes.

The most effective means of reducing mud rings is increasing the gas injection rate. A light mist often does not cut mud rings. If the gas injection rate cannot be increased due to limited capacity of the compressor or

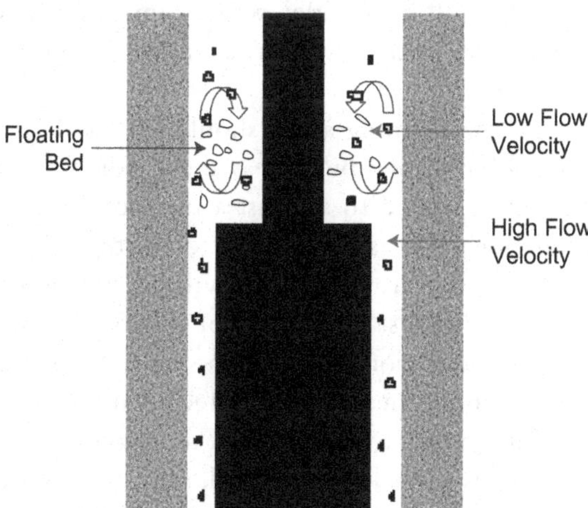

Figure 7.1 Floating bed of drill cuttings at the shoulder of drill collars.

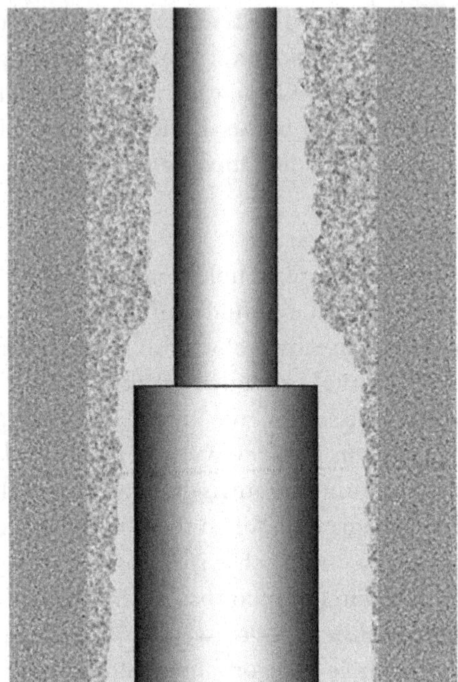

Figure 7.2 Drill cuttings above the shoulder of drill collars.

borehole washout limitations, mud rings can be cut by adding detergent to the gas. This helps solve the problem by (1) lifting the cuttings away from the area of the collar shoulder with foam and (2) softening the mud rings with surfactants that are less restrictive to the movements of the drill string.

The quantity of the foaming agent should start at 1 or 2% or about 1 or 2 quarts per bbl and then be increased according to need. Enough misting agent must be added to see just a trace of foam at the blooey line. A good starting point is 8 bbl/hour of water. Fine or dissolved cuttings, oil, salts, and water hardness will all affect misting agents to some degree. A pilot test in a quart of injection water should be performed to give an indication of where to start the mist mixture.

Tight hole problems appear to be related to mud ring problems or floating beds. It is important to not turn off the gas and to keep working the pipe. If the pipe is pulled up too hard, it may stick tight. Keeping the gas circulating is essential to preventing the growth of mud rings and reducing pipe sticking.

7.3.3 Water Removal

Liquids (water and/or oil) from wet formations accumulate at the bottomhole when the air/gas injection rate is not high enough to carry them to the surface. The accumulation of liquids increases the bottomhole pressure, which compresses the gas and reduces the gas velocity, resulting in reduced carrying capacity of the gas and, in turn, solid and more liquid accumulation at the bottomhole. This cycle will create drilling complications such as mud ringing and pipe sticking. Adding foamers (surfactants) to the gas stream can ease this problem to a certain extent. If the liquid production rate is significantly high, additional gas injection capacity is required, or the air/gas drilling needs to be converted to foam drilling. Switching to foam drilling will result in a much lower rate of penetration, and waiting for compressors with greater capacities will also reduce the overall drilling performance due to added nonrotating time. A guideline is highly desirable for drilling engineers who are making decisions about whether to convert to foam drilling.

A traditional way to determine the effect of formation fluid influx on hole cleaning is using the effective rate of penetration obtained from the equivalent rate of penetration of the influx rate (GRI, 1997; Guo and Ghalambor, 2002). Guo and colleagues (2008) conducted a comprehensive study of liquid carrying capacity of gases. They developed a systematic method for predicting the gas volume requirement necessary to remove formation fluids of various influx rates.

Starting with Turner's (Turner et al., 1969) theory of liquid loading in gas production wells, Guo and colleagues used the minimum kinetic energy criterion to establish the following theory:

$$E_{km} = 0.0576\sqrt{\sigma\rho_L} \tag{7.1}$$

where

E_{km} = minimum kinetic energy required to carry up liquid by flowing gas, lbf-ft/ft^3 or N-m/m^3

σ = interfacial tension between liquid and gas phases, dynes/cm

ρ_L = density of liquid, lb/ft^3 or kg/m^3

The typical values for water–gas interfacial tension and water density are 60 dynes/cm and 65 lbm/ft^3, respectively. Equation (7.1) yields the minimum kinetic energy value of 3.6 lbf-ft/ft^3. Since this kinetic energy value is greater than the kinetic energy value of 3 lbf-ft/ft^3 required for drill cuttings removal, this theory explains why the hole cleaning is still a problem even though the gas flow rate is high enough to remove cuttings.

The typical values for oil–gas interfacial tension and oil density are 20 dynes/cm and 45 lbm/ft^3, respectively. Equation (7.1) gives the minimum kinetic energy value of 1.73 lbf-ft/ft^3 (<3 lbf-ft/ft^3). This number implies that the required minimum gas kinetic energy in oil-influx wells is approximately half of that in water-influx wells, meaning that it is easier to clean holes with oil influx than holes with water influx. This explains why oil influx is not a significant problem in air/gas drilling.

The first solution to water removal is increasing the gas injection rate to obtain the minimum kinetic energy required to lift water. The kinetic energy per unit volume of gas can be expressed as

$$E_k = \frac{\rho_g v_g^2}{2g_c} \tag{7.2}$$

where

E_k = kinetic energy of gas, lbf-ft/ft^3 or N-m/m^3
ρ_g = density of gas, lb/ft^3 or kg/m^3
v_g = gas velocity, ft/s or m/s
g_c = 32.17 lbm-ft/lbf-s^2 or 1 kg-m/N-s^2

To evaluate the gas kinetic energy E_k in Eq. (7.2) at a given gas flow rate and compare it with the minimum required kinetic energy E_{km} in Eq. (7.1), the values of gas density ρ_g and gas velocity v_g need to be determined. The expressions for ρ_g and v_g can be obtained from the ideal gas law:

$$\rho_g = \frac{2.7 S_g p}{T} \tag{7.3}$$

and

$$v_g = 3.27 \times 10^{-2} \frac{TQ_g}{Ap} \tag{7.4}$$

where

S_g = gas specific gravity (air = 1)
p = in situ pressure, psia or kPa
T = in situ temperature, °R or °K
Q_g = gas flow rate, scf/min or scm/min
A = cross-sectional area of annular space, in^2 or m^2

Substituting Eqs. (7.3) and (7.4) into Eq. (7.2) yields

$$E_k = 4.49 \times 10^{-5} \frac{S_g T Q_g^2}{A^2 p} \tag{7.5}$$

The gas pressure p depends on the hole configuration and the gas injection rate, as shown in Eqs. (6.18), (6.26), and (6.29). For vertical boreholes, setting the left-hand side of Eq. (7.5) to be equal to E_{km} and combining it with Eq. (6.18) give the minimum gas injection rate required by water

Illustrative Example 7.1

For the following well conditions, predict the minimum required air injection rate for water removal:

Well geometry
Total measured depth: 6,000 ft
Bit diameter: 4.75 in
Drill pipe OD: 2.375 in
Material properties
Specific gravity of rock: 2.75 (water = 1)
Specific gravity of gas: 1 (air = 1)
Gas specific heat ratio: 1.25 (water = 1)
Specific gravity of oil: 1 (water = 1)
Specific gravity of water: 1.07 in
Pipe roughness: 0.0018 in
Borehole roughness: 0.1
Environment
Site elevation: 0 ft
Ambient pressure: 14.7 psia
Ambient temperature: 60°F fraction
Relative humidity: 0°F/ft
Geothermal gradient: 0.01
Operating conditions
Surface choke/flow line pressure: 14.7 psia
Rate of penetration: 60 ft/hour
Rotary speed: 50 rpm
Oil influx rate: 0 bbl/hour
Water influx rate: 8 bbl/hour
Interfacial tension: 60 dynes/cm

Solution

This problem can be solved using the chart in Figure B.10. The answer is 1,200 scf/min.

Illustrative Example 7.2

If the air injection rate of 1,200 scf/min is not available for the well in Illustrative Example 7.1, but surfactants are available to reduce the water–air interfacial tension from 60 dynes/cm to 40 dynes/cm, what is the minimum air injection rate required by water removal?

Solution

This problem can be solved using the chart in Figure B.12. The answer is 1,000 scf/min.

removal. This approach can be used to generate engineering charts for various well conditions. Some of the charts are presented in Appendix B.

The second way to remove water is by reducing the water–gas interfacial tension by adding surfactant solutions to the gas stream. This reduces the gas kinetic energy threshold required for lifting water. Various types of surfactants/foamers are available in the industry, although the cheapest surfactant is still the detergent.

7.3.4 Bit Balling

Gas drilling has the same problem with bit balling as when bits ball with mud. This happens when there is too much solids and not enough gas flow rate. Reservoirs and other low-permeability formations "weep" fluid. This leads to bit balling and mud rings. In addition to increasing the gas injection rate, adding detergent, adding a drying agent, or switching to mist can help solve these problems. Weeping often stops when the near wellbore fluids are depleted. Because they are so dry, nitrogen and natural gas are especially effective at drying a damp or weeping formation.

Another type of bit balling that is not well documented in the literature is ice balling or "frozen" bits. This was discovered when bits looked as if they were the victims of mud balling but no mud was present. The temperature of gas at the bit can be much lower than expected. This low temperature is due to the Joule–Thomson cooling effect, where a sudden gas expansion below the bit orifice causes a significant drop in temperature. The temperature can easily drop to below the ice point, resulting in ice balling of the bit if water exists. Even though the temperature can still be above the ice point, it can be below the dew point of water vapor, resulting in the formation of liquid water that causes mud ring problems in the annulus. If natural gas is used as the drilling fluid, it can form gas hydrates with water around the bit, known as hydrate balling.

Assuming an isentropic process for an ideal gas flowing through bit orifices, the temperature at the orifice downstream may be predicted using the following equation (Guo and Ghalambor, 2005):

$$T_{dn} = T_{up}\left(\frac{P_{dn}}{P_{up}}\right)^{\frac{k-1}{k}} \tag{7.6}$$

where

T_{dn} = downstream temperature, °R or °K
T_{up} = upstream temperature, °R or °K
P_{dn} = downstream pressure, lb/ft^2 or N/m^2
P_{up} = upstream pressure, lb/ft^2 or N/m^2

The upstream temperature may be lower than the geothermal temperature at the bit depth because the downstream gas cools the bit body, and the bit body, in turn, cools the upstream gas. The process can continue until a dynamic equilibrium with geothermal and gas temperatures is reached at the bottom of the hole.

The downstream temperature predicted by Eq. (7.6) should be compared with the ice point (the hydrate point for gas drilling with natural gas) and the dew point of the water at in situ pressure to identify possible ice/hydrate and condensation problems. The ice point of water at bottomhole pressure may be assumed to be 32°F. The hydrate point of water can be found in many sources, including Guo and Ghalambor (2005). The dew point of water vapor at bottomhole pressure can be estimated based on the humidity of the gas, the water removal efficiency of the compressor, the in situ water saturation pressure, and the pressure above the bit orifice.

Two solutions to the ice/hydrate balling problem are reducing the gas injection rate to avoid a sonic flow (see the next section) and using bits with large orifices without installing bit nozzles. Both of these methods can reduce the pressure drop at the bit so the downstream to upstream pressure ratio will not become too low.

7.3.5 Problems Associated with Sonic Flow at the Bit

Two flow conditions required for fluid to flow through restrictions such as orifices and nozzles are sonic flow and subsonic flow, also referred to as critical flow and subcritical flow. Pressure waves, being mechanical waves, obey the same principle as sound waves. When the fluid flow velocity in an orifice is equal to or greater than the traveling velocity of sound in the fluid under the in situ condition, the flow is called a sonic flow. Under a sonic flow condition, the pressure wave downstream of the orifice cannot propagate upstream through the orifice because the medium (fluid) is traveling in the opposite direction at the same or higher velocity. This causes a pressure discontinuity at the orifice—that is, the upstream pressure is not influenced by the downstream pressure.

A sonic flow can have many harmful effects on gas drilling operations, including pipes sticking, ice/hydrate balling of bits, wellbore washouts, and crooked holes. Because of the pressure discontinuity at the orifice, any increase in bottomhole pressure due to cuttings accumulation or mud ringing in the annulus cannot be detected by reading the standpipe pressure gauge. Cuttings will continue to accumulate until the drill string gets stuck. Often pipe sticking occurs only a few minutes after a "smooth" drilling operation. The operation looked smooth because the standpipe pressure was normal, while the annular pressure had already increased due to cuttings accumulation or mud ringing. To reduce the possibility of pipe sticking, sonic flow should be avoided by using larger bit nozzles or orifices in all gas drilling operations.

Whether sonic flow exists at the bit depends on the downstream–upstream pressure ratio. If this pressure ratio is less than a critical pressure ratio, sonic flow exists. The critical pressure ratio depends on the fluid properties, not on the configuration of the orifice. It is expressed as

$$\left(\frac{P_{dn}}{P_{up}}\right)_c = \left(\frac{2}{k+1}\right)^{\frac{k}{k-1}} \tag{7.7}$$

Since the values of the specific heat ratio (k) of air and natural gases are between 1.2 and 1.4, Eq. (7.7) gives the critical pressure ratio values ranging from 0.51 to 0.53. Use of these numbers in Eq. (7.6) shows an 84% reduction in the absolute temperature scale and more reduction in the relative temperature scale (°F or °C).

Under sonic flow conditions the upstream pressure is given by the following equation for ideal gases:

$$P_{up} = \frac{Q_g}{609.33A_o}\sqrt{\frac{S_g T_{up}}{k\left(\frac{2}{k+1}\right)^{\frac{k+1}{k-1}}}} \tag{7.8}$$

where

Q_g = gas flow rate, scf/min or scm/min
A_o = total area of bit orifices, in^2 or m^2

The flow equation for subsonic flow is Eq. (6.30):

$$Q_g = 6.02CA_oP_{up}\sqrt{\frac{k}{(k-1)S_gT_{up}}\left[\left(\frac{P_{dn}}{P_{up}}\right)^{\frac{2}{k}}-\left(\frac{P_{dn}}{P_{up}}\right)^{\frac{k+1}{k}}\right]} \quad (7.9)$$

Substituting $C = 0.6$, $k = 1.4$, and the critical pressure ratio of 0.53 into this equation gives an expression of maximum gas flow rate without causing sonic flow as

$$Q_{gmax} = 1.75\frac{A_oP_{up}}{\sqrt{S_gT_{up}}} \quad (7.10)$$

If the operating gas injection rate is higher than this value, larger orifice area A_o should be utilized.

The problem of wellbore washout is frequently encountered in gas drilling. It may be attributed to three effects: High-velocity gas out of the bit orifice can create hole enlargement in soft formations at the bottomhole;

Illustrative Example 7.3

For the well conditions given in Illustrative Example 6.4, check the design gas injection rate of 1,615 scf/min against the maximum gas flow rate without causing sonic flow.

Solution

The computer program *GasDrill-09.xls*.gives the follow data:

Total bit orifice area: 0.92 in^2
Upstream pressure of bit orifice: 15,845 lb/ft^2
True vertical depth at collar shoulder: 4,327 ft
Temperature at collar shoulder: 578°R

Equation (7.10) gives

$$Q_{gmax} = 1.75\frac{(0.92)(15{,}845)}{\sqrt{(1)(578)}} = 2{,}345 \text{ scf/min}$$

The design gas injection rate of 1,615 scf/min is less than 2,345 scf/min, so sonic flow is not expected.

cold gas from the bit orifice can cause failure of the borehole wall due to local thermal stress at the bottomhole; and high-velocity gas in the upper hole sections can cause borehole erosion in soft formation intervals. All of these effects can be minimized by reducing the gas injection rate. If this is not an option due to hole cleaning concerns, using large bit orifices can remedy the first two problems. A flow diverging joint (FDJ) can be installed at the drill collar shoulder to reduce gas flow through the bit without affecting hole cleaning in the borehole above it. For the third problem, sometimes it is necessary to set the casing deeper to protect soft formations from erosion.

A crooked hole is usually not a problem in gas drilling operations when air hammers with flat-bottom bits are used. Air hammers require a very low weight on bit (WOB) to drill even very hard formations. A low WOB allows for a straight bottomhole assembly (BHA) while drilling, resulting in straight holes. However, crooked holes are occasionally reported from gas drilling operations. This usually occurs when conventional rock bits are used along with high injection rates of gases. In these situations, bottomhole washout is believed to be responsible for hole deviations. Reducing the gas injection rate and/or using large bit orifices should ease the problem. Use of FDJ at the drill collar shoulder is another option.

7.3.6 Corrosion

Corrosion occurs in wet systems, such as mists and foams. If misting is implemented for lifting large volumes of water, the misting agent needs to be added in a much higher concentration. The pH and corrosion control must be considered on a case-by-case basis. pH control is important to avoid corrosion in air drilling and in operations using a nitrogen membrane system (which introduces small levels of oxygen downhole). The pH of the mixture must be kept above 9 at the blooey line. Lime and cement have been used in injection water, but they both leave scale and can damage foamers. On the other hand, the presence of calcium ions reduces shale swelling. KCl appears to be a better choice, since it does not leave scale in the pipe and it has a buffer effect. NaOH or caustic soda is not buffered enough to maintain its pH. Soda ash (Na_2CO_3) has been used in some areas.

Corrosion inhibitors with air drilling or in the presence of oxygen from a nitrogen membrane include phosphates and filming amines. Different compounds have been found to work satisfactorily in various areas. Corrosion in freshwater appears to be minimized, although it is worse when using saline water. With produced water, corrosion can be significant, and in produced water with traces of H_2S, corrosion is very difficult to control. There have been local studies of corrosion control, but there is no single study that covers worldwide operations or that profiles any significant direction. Corrosion control, like misting agents, is a local issue.

Corrosion inhibition, misting, and foaming affect one another in a negative way. Increasing the conductivity of the electrolyte in a system (the fluid) will cause corrosion to increase. Brine water will increase the conductivity and will act to destabilize the foam. As the foam begins to break down, the amount of free water will increase. This further accelerates the corrosion rate. As temperature increases, corrosion increases. Foams exposed to higher temperatures also begin to break down, yielding additional free water. This compounds the rate of corrosion. Oxygen is probably the worst offender to foam with respect to corrosion. Any foam system generated with air will be potentially corrosive if it is not well controlled. The two types of corrosion inhibitors are anionic inhibitors, which are more compatible with foaming agents, and cationic inhibitors, which tend to act contrary to the foaming agent and have a destabilizing effect on the fluid system.

7.3.7 Safety

Safety issues in gas drilling mainly involve hydrogen sulfide (H_2S) natural gas, which is highly toxic and life threatening. Natural gas, either injected as a drilling fluid or produced from a formation, can cause fires and explosions at the surface if it is not handled correctly. Natural gas from formations can also cause downhole fires and explosions when air is used as the drilling fluid.

Hallman and colleagues (2007) reported on a case involving H_2S operations. Gas drilling operations where H_2S may be present must include warnings of hazards and the following measures:

- Adequate crew training
- Special safety equipment (sensors, alarms, respirators, etc.)
- Emergency contingency plan
- H_2S-resistant materials and training
- Pressured surface separation vessels
- Auxiliary vacuum degassing equipment

Contingency plans must be carefully developed before the drilling operations begin. Casing programs, circulation designs, and onsite quality control and monitoring are particularly important. Operational and equipment testing procedures must be enforced and be well understood by all personnel.

The importance of safety equipment can never be overemphasized in H_2S operations. Remember:

- Operational and equipment testing procedures must be established, comprehended by all personnel, and enforced.
- Drilling should not continue if pressures exceed the maximum limits established.
- Emphasis is placed on monitoring pressure while drilling, tripping, and stripping.
- A blowout preventer (BOP) stack must be tested each time it is reinstalled.
- Surface equipment should be regularly inspected and monitored.
- If H_2S is detected, stop drilling.
- Inspect liquid/gas separators daily.
- Inspect diverter rubber elements several times a day.
- Check the diverter alignment with the rotary table.
- Have a contingency plan.

To prevent fires or explosions of natural gas at the surface, flaring gas is a must.

- Flare lines should be adequately sized.
- Flare stacks should be properly positioned.
- Use of automatic flame igniters is preferred.
- Wind direction should be considered.
- Flare stack height should be adjusted for optimum performance.
- Flare lines should be adequately anchored.

When drilling with natural gas, liquid hydrocarbon separation and storage facilities must meet API RP 500B, National Fire Protection Association (NFPA) 70, and NFPA 496 guidelines.

Using float equipment is always a good practice in gas drilling to prevent backflow. For optimum conditions, a good rule of thumb is to install a float every 12 joints. Two floats should be placed close to the surface to minimize the time required to bleed off pressure before making a connection.

Downhole fires and explosions can occur when hydrocarbons from the formation are mixed with oxygen at high temperatures that result from the mechanical friction between the drill string and the borehole with a mud ring. Extreme caution must be exercised when air is used as a drilling fluid. Although flammable conditions are well established in terms of natural gas content in a mixture and the minimum oxygen required for a flammable mixture (GRI, 1997), these facts are seldom considered when designing air drilling operations. This is mostly because downhole conditions are very hard to predict. The best practice is to avoid using air as a drilling fluid whenever hydrocarbon-bearing zones are drilled. If air has to be utilized, use misting/foaming to reduce mud rings and thus the potential for fires or explosions.

Thorough training of the drilling crew is essential for safe gas drilling operations. Personnel safety training and detailed, written gas drilling procedures are required. An emergency backup escape route is imperative in the event that the wind changes direction. Gas detection, fire extinguishing, and other equipment should be placed at strategic locations.

SUMMARY

This chapter introduced operating procedures in gas drilling to reduce operation complications and outlined measures to avoid gas drilling problems. Possible problems include borehole instability, mud rings, water loading, bit balling, crooked holes, corrosion, and downhole fires and explosions. An inadequate gas injection rate is recognized as the leading cause of all of these problems. Overinjection of gas and small bit orifices are believed to be responsible for bit balling with ice or hydrates, wellbore washouts, and crooked holes. Thorough training of the drilling crew is essential for safe gas drilling operations.

REFERENCES

Gas Research Institute (GRI), 1997. Underbalanced Drilling Manual. Gas Research Institute Publication, Chicago.

Guo, B., Ghalambor, A., 2002. Gas Volume Requirements for Underbalanced Drilling of Deviated Holes. PennWell Books.

Guo, B., Ghalambor, A., 2005. Natural Gas Engineering Handbook. Gulf Publishing Company.

Guo, B., Yao, Y., Ai, C., 2008. Liquid carrying capacity of gas in underbalanced drilling. Paper SPE 113972, Proceedings of the SPE Western Regional Meeting, March 31–April 2, Los Angeles.

Hallman, J.H., Cook, I., Muqeem, M.A., Jarrett, C.M., Shammari, H.A., 2007. Fluid customization and equipment optimization enable safe and successful underbalanced drilling of high-H2S horizontal wells in Saudi Arabia. Paper SPE 108332, Proceedings of the IADC/SPE Managed Pressure Drilling and Underbalanced Operations Conference and Exhibition, March 28–29, Galveston.

Turner, R.G., Hubbard, M.G., Dukler, A.E., 1969. Analysis and prediction of minimum flow rate for the continuous removal of liquids from gas wells. J. Pet. Technol., Trans. AIME 246, 1475–1482.

PROBLEMS

7.1 Solve Illustrative Example 7.1 assuming water influx is 12 bbl/hour.

7.2 Solve Illustrative Example 7.1 assuming water–gas interfacial tension of 40 dynes/cm.

7.3 For the well conditions given in Illustrative Example 6.4, predict the gas temperature downstream of the bit orifice.

PART THREE

Underbalanced Drilling Systems

Underbalanced drilling is used mainly for drilling oil and gas recovery wells. Liquid foam, aerated liquid, and oil are the popular fluids used in unbalanced drilling (UBD). Since these fluids have densities lower than water, they are utilized for drilling oil and gas pay zones with subnormal or depleted pressures. The main purpose of UBD is to protect these pay zones from being damaged by drilling fluids.

This part of the book provides drilling engineers with basic information about UBD systems and the techniques used to improve UBD performance. Materials are presented in three chapters:

Chapter 8: Equipment in Underbalanced Drilling Systems
Chapter 9: Gas and Liquid Injection Rates
Chapter 10: Underbalanced Drilling Operations

CHAPTER EIGHT

Equipment in Underbalanced Drilling Systems

Contents

8.1 INTRODUCTION

Liquid foam, aerated liquid, and oil are used as the circulating fluids in underbalanced drilling (UBD) operations. The same types of equipment are used in UBD with oil as for normal liquid drilling, except more reliable well-control equipment is emphasized. The equipment in foam and aerated liquid UBD systems includes gas compressors and separators in addition to the liquid drilling systems. This chapter provides a brief introduction to the equipment used in foam and aerated liquid UBD systems.

8.2 SURFACE EQUIPMENT

Figure 8.1 shows a layout of surface equipment in a closed system using aerated liquid as the drilling fluid. The liquid pump is usually the same pump as for mud drilling. The nitrogen pumpers are gas compressors. The nitrogen gas is usually obtained using nitrogen generators, as shown in Figure 5.4. Liquid nitrogen has been employed in offshore UBD operations. A 4-phase separator (Figure 8.2) separates gas, oil, drilling fluid, and drilling solids. The rotating head used in gas drilling is not popular in UBD operations due to its low pressure rating. Rotating blowout preventers (BOPs) are often used in UBD (Figure 8.3) (GRI, 1997). They can handle up to 2,500-psi wellhead pressure.

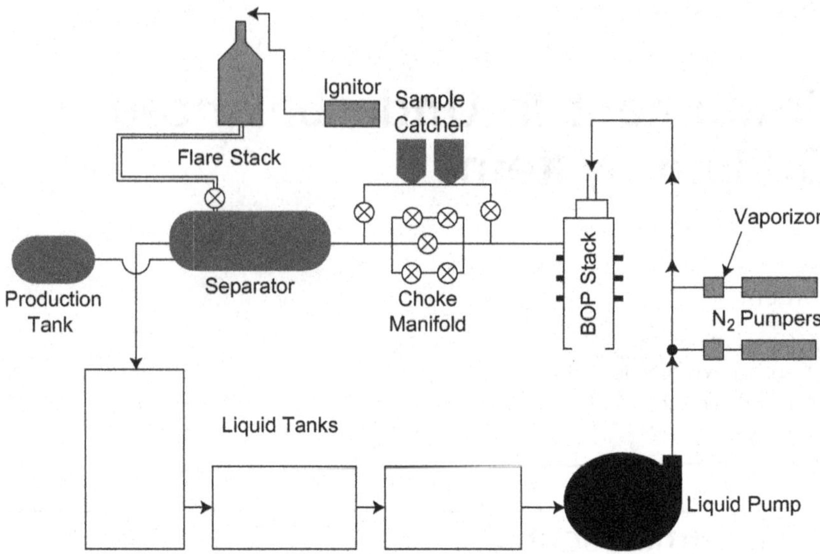

Figure 8.1 A closed underbalanced drilling system.

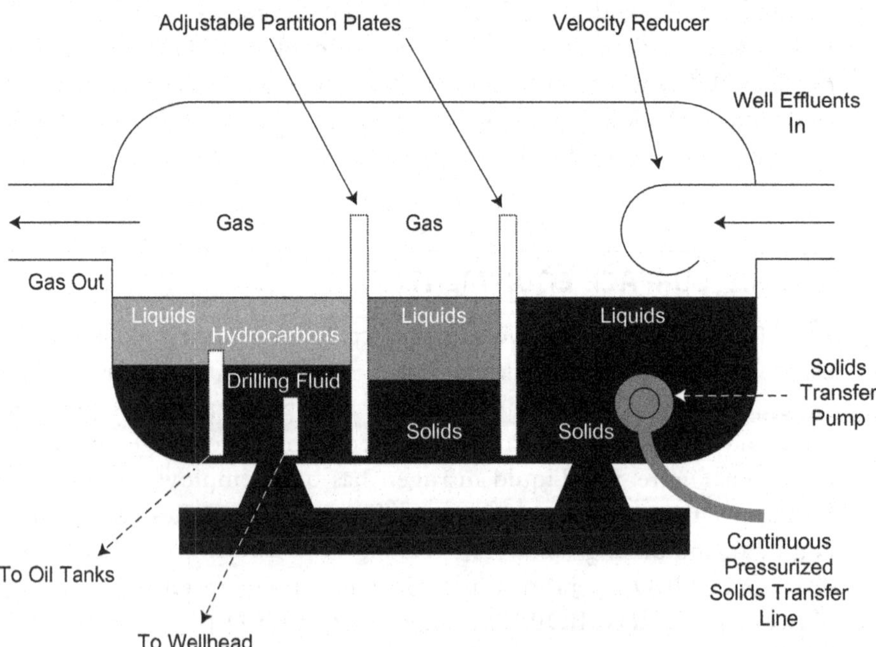

Figure 8.2 A 4-phase separator used in underbalanced drilling operations.

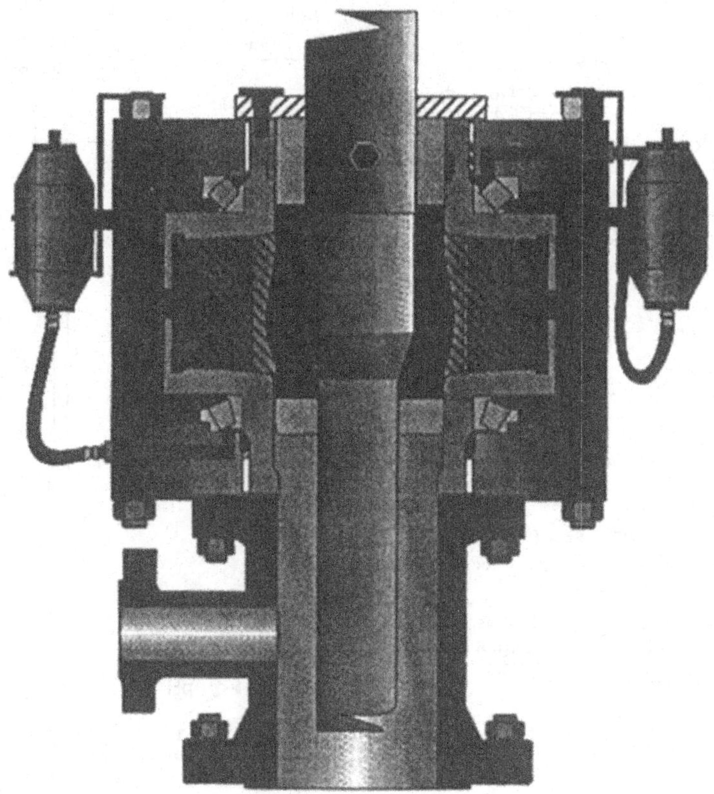

Figure 8.3 A cutoff view of a rotating blowout preventer.

In foam UBD operations, the foam returned from the hole can be either destroyed and discarded or recycled. Figure 8.4 shows a foam separation system for disposal. The returned foam is destroyed by injecting foam killer (acids) into the blooey line and is then separated by a cyclone. Figure 8.5 shows a foam recycling system. The returned foam is destroyed by feeding it with foam killer (acids) in the blooey line. After separation, about 95% of foaming agents remain in the liquid phase. By adding alkali to raise its pH value, the liquid will form foam with gas again.

Snubbing units are employed in some underbalanced drilling operations. They are similar in principle to those used in well workover operations (Rehm, 2002).

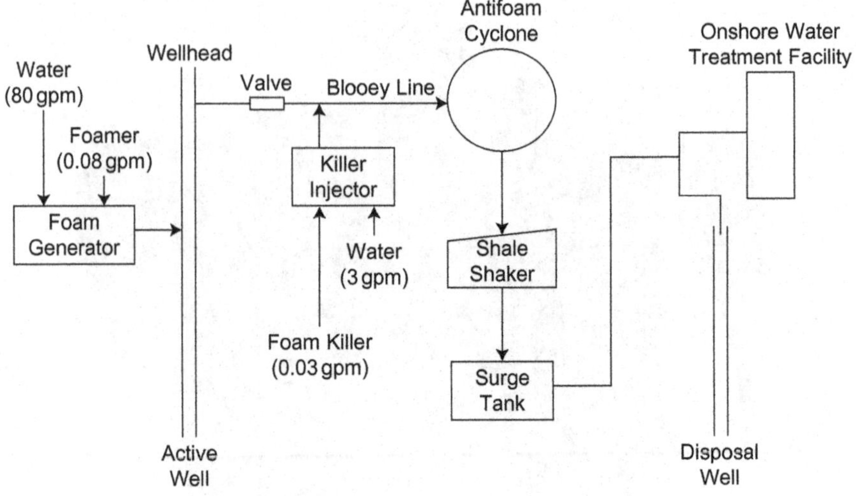

Figure 8.4 A foam separation system.

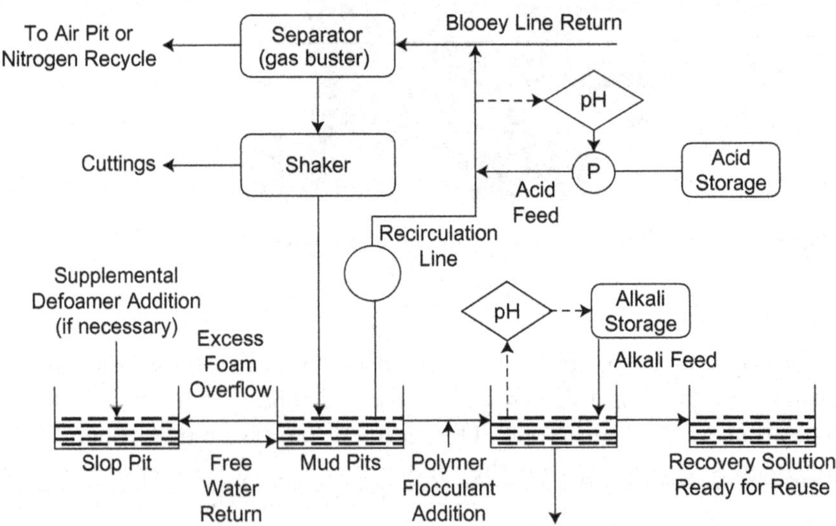

Figure 8.5 A foam recycling system.

8.3 DOWNHOLE EQUIPMENT

Like surface equipment, some downhole equipment in liquid and gas drilling operations—for example, some types of drill pipe valves—are also used in UBD operations. Mud motors in liquid drilling are also used

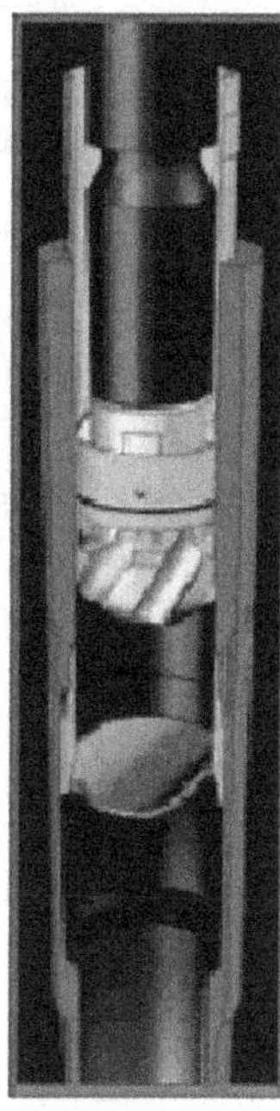

Figure 8.6 Halliburton's deployment valve, the Quick-Trip® Valve. *(Courtesy of Halliburton.)*

in UBD for horizontal drilling. Some unique types of equipment that are widely used in UBD are the deployment valves for underbalanced trip operations. Both mechanically and hydraulically actuated valves are available in the oil and gas industry. Figure 8.6 shows a cutoff view of a mechanically controlled deployment valve.

The deployment valve isolates the reservoir while tripping during drilling operations. It establishes a mechanical barrier, eliminating the need to kill the well. Mechanical actuation eliminates external control lines. It enables the safe deployment of slotted liners or production screens without killing the well. It can be a permanent or temporary installation. Deployment valves make the UBD operations easier and safer.

SUMMARY

This chapter provided a brief introduction to some special types of equipment that are used in UBD operations. Key equipment includes blowout preventers, 4-phase separators, and downhole deployment valves.

REFERENCES

GRI, 1997. Underbalanced Drilling Manual. Gas Research Institute Publication, Chicago.
Rehm, B., 2002. Practical Underbalanced Drilling and Workover. University of Texas at Austin.

PROBLEMS

8.1 How does the pH value of drilling liquid affect the performance of foaming agents?
8.2 What is the working principle of 4-phase separators?
8.3 Why do downhole deployment valves make UBD operations safer and more cost effective?

CHAPTER NINE

Gas and Liquid Injection Rates

Contents

9.1 INTRODUCTION

Most underbalanced drilling operations are performed using aerated liquids or foams. Injecting the right amounts of gas and liquid into the well is crucial for the success of unbalanced drilling (UBD) operations. This is because the injection rates determine the borehole pressure that is responsible for formation damage and borehole damage (borehole collapses and washouts). The injection rates also affect hole cleaning. This chapter presents multiphase flow models and their applications to designing gas and liquid injection rates for UBD operations using aerated liquids and foams as drilling fluids.

9.2 MULTIPHASE FLOWS IN UBD SYSTEMS

Multiphase fluids used in UBD operations include aerated liquids and foams. These fluid systems consist of gas (normally air or nitrogen), liquid (normally water or oil), and solid (cuttings). The liquid constitutes the continuous phase, with gas appearing as discontinuous bubbles. Both steady-state and transient flow models are available for multiphase flow drilling hydraulics calculations. Unfortunately, the results from these models frequently conflict because of assumptions made in the mathematical formulas (Nakagawa, 1999). Steady-state flow models are widely used for designing multiphase flow hydraulics programs, while transient flow models are often employed for extreme case studies.

9.2.1 Flow Regimes

Multiphase flow is much more complicated than single-phase flow due to the variation of flow regimes (or flow patterns). Fluid distribution changes greatly in different flow regimes, which significantly affects pressure gradients inside and outside the drill string. As shown in Figure 9.1 (Govier and Aziz, 1977), at least five flow regimes have been identified for gas-liquid two-phase flow in vertical conduits: bubble, slug, churn, annular, and mist flow. These flow regimes occur as a progression with an increasing gas flow rate for a given liquid flow rate. The former three flow regimes are often observed in UBD operations, while the latter two are most often encountered in gas drilling operations.

In a bubble flow, the gas phase is dispersed in the form of small bubbles in a continuous liquid phase. In a slug flow, gas bubbles coalesce into larger bubbles that eventually fill the entire pipe cross-section. Between the large bubbles are slugs of liquid that contain smaller bubbles of entrained gas. In a churn flow, the larger gas bubbles become unstable and collapse, resulting in a highly turbulent flow pattern with both phases dispersed. In an annular flow, gas becomes the continuous phase, with liquid flowing in an annulus coating the surface of the pipe and with droplets entrained in the gas phase. In a mist flow, liquid is entrained in the continuous gas phase in the form of mist.

9.2.2 Liquid Holdups

The liquid phase always flows slower than the gas phase in upward flow streams and faster in downward flow streams. The differences in phase velocities cause the in situ volume fractions of fluids to be different from the volume fractions at the injection point (surface). To be more specific, the amount of pipe occupied by a phase is often different from its proportion of the total volumetric flow rate. This is due to the differences in density between the phases. Gravity causes the dense phase to slip down in an upward flow—that is, the lighter phase moves upward faster than the denser phase.

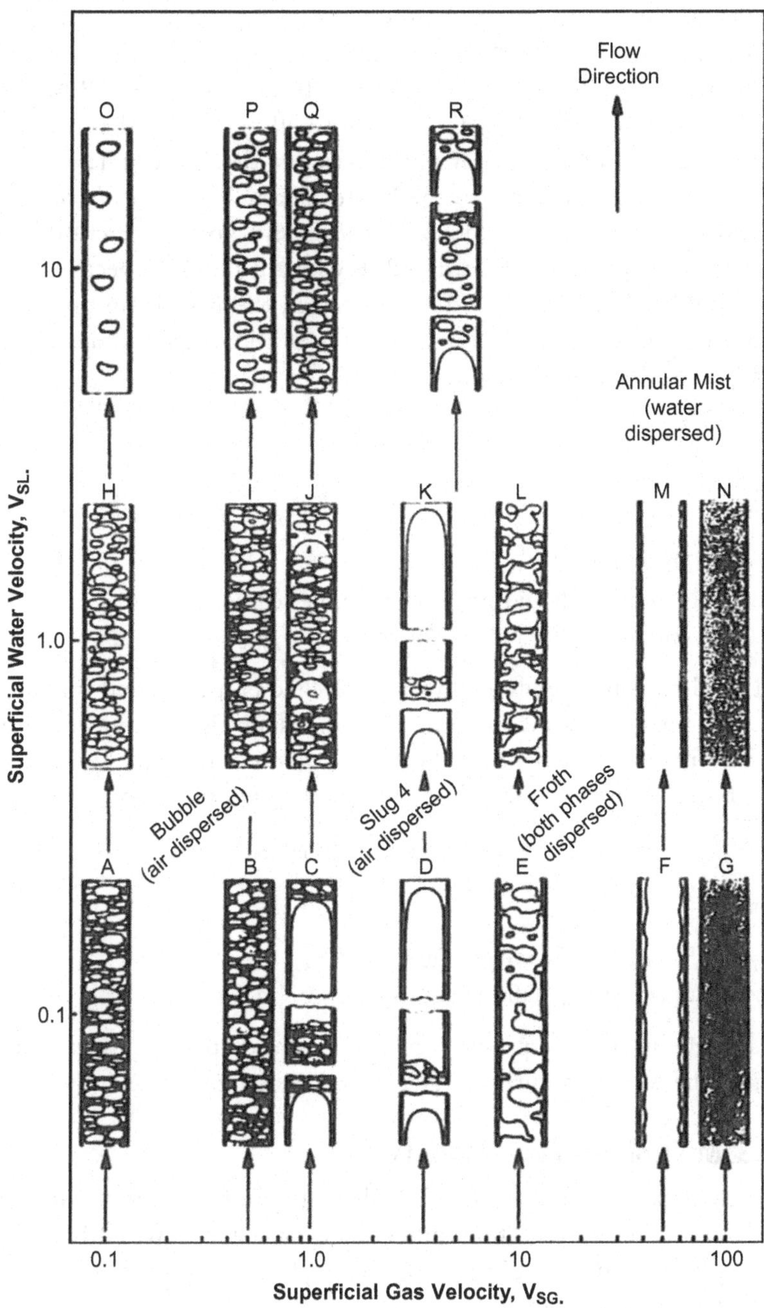

Figure 9.1 Flow regimes in a gas–liquid flow (Govier and Aziz, 1977).

On the contrary, gravity causes the dense phase to slip down in a downward flow—that is, the lighter phase moves downward slower than the denser phase. Because of the gravity effect, in an upward flow the in situ volume fraction of the denser phase will be greater than the input volume fraction of the denser phase—that is, the denser phase is "held up" in the conduit relative to the lighter phase. The terms *liquid holdup* and *gas holdup* are used to describe the in situ volume fraction of the liquid and gas phase in upward and downward flows, respectively. The liquid holdup has been well documented in the petroleum production literature, while the gas holdup has not been thoroughly studied. Liquid holdup is mathematically defined as

$$y_L = \frac{V_L}{V} \tag{9.1}$$

where

y_L = liquid holdup, fraction
V_L = volume of liquid phase in the pipe segment (ft^3, m^3)
V = volume of the pipe segment (ft^3, m^3)

Liquid holdup depends on flow regime, fluid properties, and conduit size and configuration. Its value can only be quantitatively determined through experimental measurements. A direct application of the liquid holdup is to use it for estimating mixture specific weight in a two-phase flow:

$$\gamma_{mix} = y_L \gamma_L + (1 - y_L)\gamma_G \tag{9.2}$$

where

γ_L = liquid specific weight (lb/ft^3, N/m^3)
γ_G = in situ gas specific weight (lb/ft^3, N/m^3)

Because the in situ gas specific weight is much less than the liquid specific weight, the former is usually neglected in most engineering analyses.

9.2.3 Multiphase Flow Models

Mathematical models used for describing multiphase flow fall into two categories: homogeneous flow models and separated flow models. Liquid holdup is not considered in homogeneous flow models but is considered in separated flow models.

Bubbly flow regimes can be approximately described by homogeneous flow models. Lage and Time's (2000) work indicates that a bubbly flow exists when the in situ gas–liquid ratio (GLR, dimensionless) is less than unity. It also shows that a dispersed bubble flow occurs for superficial liquid velocities greater than 6 ft/sec and superficial gas velocities as high as 12 ft/sec. The research work by Sunthankar and colleagues (2001) on multiphase flow in an inclined well model confirmed Lage and Time's findings that a bubbly flow exists in the annular space when the in situ GLR is less than unity. It can be shown that the in situ GLR is greater than 1 only in a small portion of borehole sections (near the surface) in the aerated liquid drilling practice (EMW between 4.0 and 6.9).

Although separated flow models are believed to be more accurate than homogeneous flow models, the latter is still attractive and is widely used due to its simplicity. In fact, it has been shown that the homogeneous flow models are accurate enough in UBD hydraulics calculations (Guo et al., 1996; Guo et al., 2003; Sun et al., 2004).

Guo and Colleagues' Homogeneous Flow Model

Guo and colleagues (1996) developed their first homogeneous flow model using numerical integration. The model was validated with field data from three wells at various depths. The model can simulate conventional aeration, jet sub injection, and parasite tubing injection. In 2003, Guo and colleagues presented a closed form hydraulics equation for predicting bottomhole pressure in UBD with foam. In 2004, Guo, Sun, and colleagues published a closed form hydraulics equation for aerated mud drilling in inclined wells. For simplicity, only the closed form models are included in this section.

Aerated Liquid Drilling Models

Guo, Sun, and colleagues' aerated liquid drilling model is capable of simulating gas, water, oil, and solid 4-phase flows. The model takes the following form:

$$b(P-P_s)+\frac{1-2bm}{2}\ln\left|\frac{(P+m)^2+n}{(P_s+m)^2+n}\right| - \frac{m+bn-bm^2}{\sqrt{n}}\left[\tan^{-1}\left(\frac{P+m}{\sqrt{n}}\right)-\tan^{-1}\left(\frac{P_s+m}{\sqrt{n}}\right)\right] = a(1+d^2e)L \tag{9.3}$$

where P is pressure in lbf/ft^2 at conduit length L in ft; P_s is pressure in lbf/ft^2 at conduit length $L = 0$; and

$$a = \frac{0.0014 d_b{}^2 S_s R_p + 0.25 W_m Q_m + 1.44 S_l Q_f + 0.019 S_g Q_{go}}{T Q_{go}} \cos(\theta) \quad (9.4)$$

$$b = \frac{0.033 Q_m + 0.023 Q_f}{T Q_{go}} \quad (9.5)$$

$$c = \frac{9.77 T Q_{go}}{A} \quad (9.6)$$

$$d = \frac{0.33 Q_m + 0.22 Q_f}{A} \quad (9.7)$$

$$e = \pm \frac{f}{2 g D_H \cos(\theta)} \quad (9.8)$$

$$m = \frac{cde}{1 + d^2 e} \quad (9.9)$$

$$n = \frac{c^2 e}{(1 + d^2 e)^2} \quad (9.10)$$

where

g = 32.2 ft/sec^2; d_b = bit diameter, in
S_s = solid specific gravity, water = 1
R_p = rate of penetration (ROP), ft/hr
W_m = liquid weight, lbm/gal
Q_m = liquid flow rate delivered by pump, gal/min
S_l = formation fluid specific gravity, water = 1
Q_f = formation fluid influx rate, bbl/hr
Q_{go} = volumetric gas flow rate, ft^3/min
S_g = specific gravity of gas, air = 1
T = absolute temperature, °R
θ = inclination angle, degrees
A = cross-sectional area of flow path, in^2
D_H = hydraulic diameter of the flow path, ft
f = friction factor

The positive and negative signs in Eq. (9.8) are the upward and downward flows, respectively.

Determining the friction factor for multiphase flows presents a major challenge in hydraulics calculations. Although a number of friction factor correlations have been used by previous investigators (Caetano et al., 1992; Nakagawa et al., 1999; Lage and Time, 2000; Lyons et al., 2001), their accuracies are debatable.

For aerated liquid flow, Guo, Sun, and colleagues proposed the following friction factor expression:

$$f = F_{LHU}\left[\frac{1}{1.74 - 2\log\left(\dfrac{2\bar{e}}{D_H}\right)}\right]^2 \tag{9.11}$$

in which $\bar{e}$ = the average wall roughness (0.00015 ft for steel pipes and 0.004 ft for openhole walls), and F_{LHU} = a correction factor accounting for liquid holdup in multiphase flows. Guo, Sun, and colleagues used the borehole pressure measurements at Petrobras's Research and Training Facility in Taqyuipe, Bahia (Nakagawa et al., 1999; Lage and Time, 2000; Lage et al., 2000), to correlate F_{LHU} to the average GLR downstream of the point of interest. The F_{LHU} was determined to be

$$F_{LHU} = (13.452 - 0.02992G_{LR})/F_t \tag{9.12}$$

where

G_{LR} = average downstream GLR (dimensionless)
F_t = tuning factor ($F_t \approx 2$)

The G_{LR} can be estimated with the following relation:

$$G_{LR} = \frac{14.7Q_{go}}{\left[\dfrac{P_s + P}{(2)(144)}\right]\left(\dfrac{Q_m}{7.48} + \dfrac{5.615Q_f}{60}\right)} \tag{9.13}$$

Because the G_{LR} depends on the pressure at the point of interest, Eq. (9.13) should be implicitly involved in the numerical procedure for pressure calculations.

The multiphase pressure drop at the bit can be expressed as (Guo et al., 1996)

$$\Delta P_b = \frac{\dot{W}_t^2}{gA_n^2}\left(\frac{1}{\gamma_{dn}} - \frac{1}{\gamma_{up}}\right) \tag{9.14}$$

where

ΔP_b = pressure drop at bit (lbf/ft^2)
$\dot{W}_t$ = total weight flow rate (lbf/s)
A_n = total bit nozzle area (ft^2)
γ_{dn} = specific weight of mixture at downstream (lbf/ft^3)
γ_{up} = specific weight of mixture at upstream (lbf/ft^3)

This equation is valid for all multiphase flow, including aerated liquids and foams.

Illustrative Example 9.1

A well is to be cased to 5,291 ft with a 8⅝-in, 28-lb/ft (8.017-in ID) casing. Starting from the kickoff point at 5,321 ft, the hole will be drilled with mud using a 7⅞-in bit to build an inclination angle at a constant build rate of 4°/100 ft until the maximum inclination angle of 90 degrees is reached at a depth of 7,563 ft. Then the drilling fluid will be shifted from mud to aerated liquid to drill to the TD of 8,050 ft while the inclination angle is maintained at 90 degrees. Additional data are given as follows. Calculate and plot profiles of pressure, velocity, GLR, mixture density, and ECD along the flow path when the bit is at the total depth.

Drill string data
Length of heavy drill pipe below collar: 1,500 ft
Heavy drill pipe OD: 5 in
Heavy drill pipe ID: 3 in
Drill collar length: 480 ft
Drill collar OD: 6.25 in
Drill collar ID: 2.5 in
Drill pipe OD: 4.5 in
Drill pipe ID: 3.643 in
Material properties
Specific gravity of rock: 2.65 (water = 1)
Specific gravity of gas: 1 (air = 1)
Density of injected liquid: 8.4 ppg
Specific gravity of formation water: 1.05 (water = 1)
Specific gravity of formation oil: 0.85 (water = 1)

Pipe roughness: 0.0018 in
Borehole roughness: 0.1 in

Environment

Site elevation (above mean sea level): 50 ft
Ambient pressure: 14.7 psia
Ambient temperature: 60 F
Relative humidity: 0.1 fraction
Geothermal gradient: 0.01 F/ft
Friction tuning factor: 8

Operating conditions

Surface choke/flow line pressure: 14.7 psia
Rate of penetration: 120 ft/hour
Gas injection rate: 560 scfm
Liquid injection rate: 275 gpm
Formation water influx rate: 1 bbl/hour
Formation oil influx rate: 1 bbl/hour
Bit orifices: 20-1/32nd in
20-1/32nd in
20-1/32nd in

Solution

This problem can be solved with computer program *AliqDrill-09.xls*. The results are shown in Figures 9.2 through 9.6.

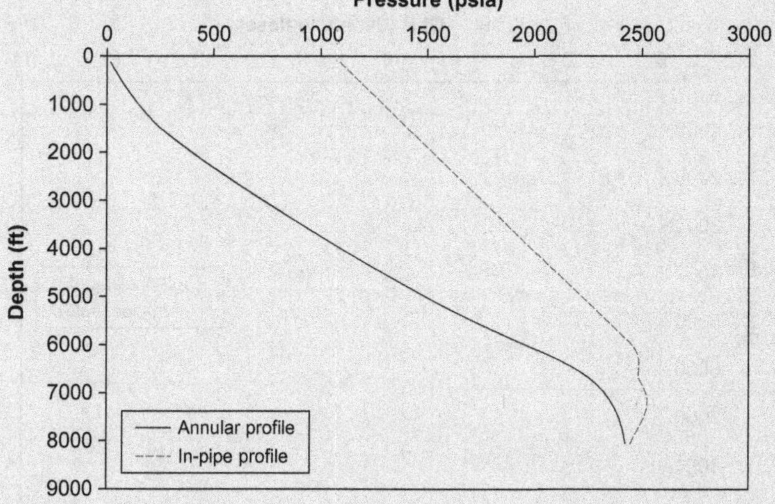

Figure 9.2 Calculated pressure profile along the flow path.

(Continued)

Illustrative Example 9.1 *(Continued)*

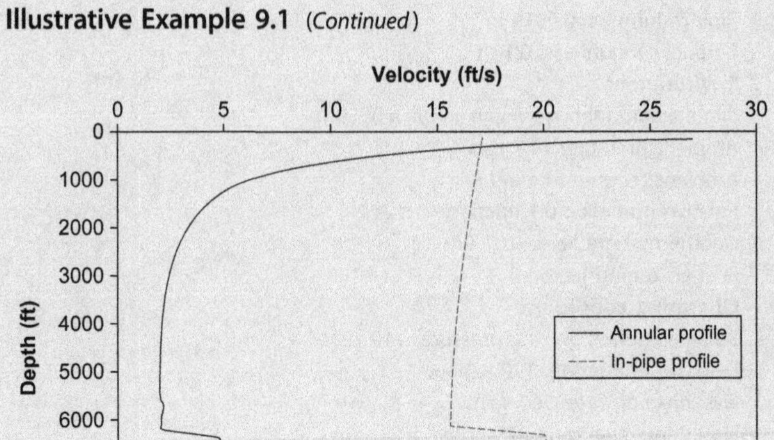

Figure 9.3 Calculated velocity profile along the flow path.

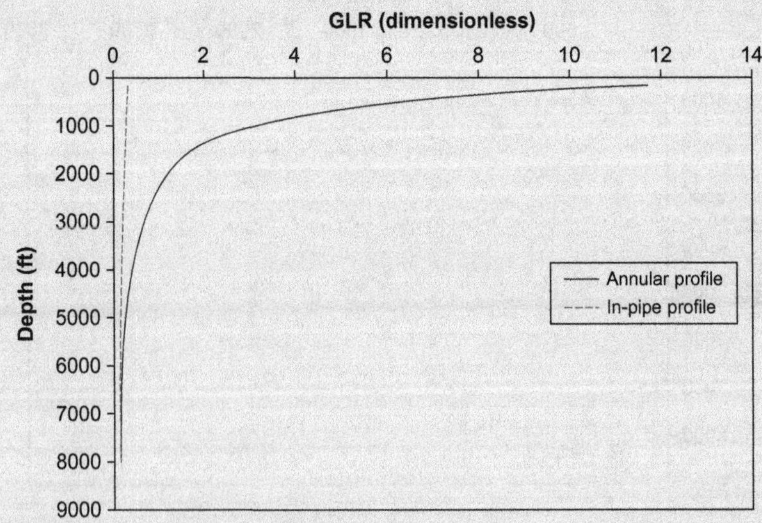

Figure 9.4 Calculated GLR profile along the flow path.

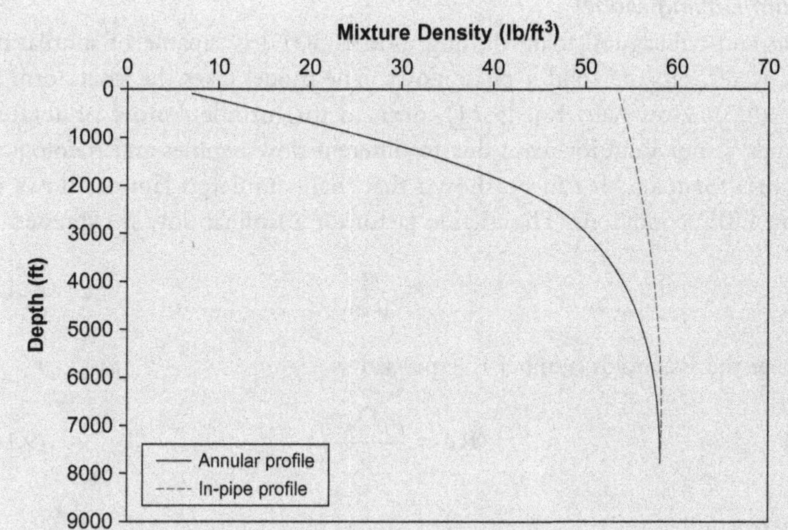

Figure 9.5 Calculated density profile along the flow path.

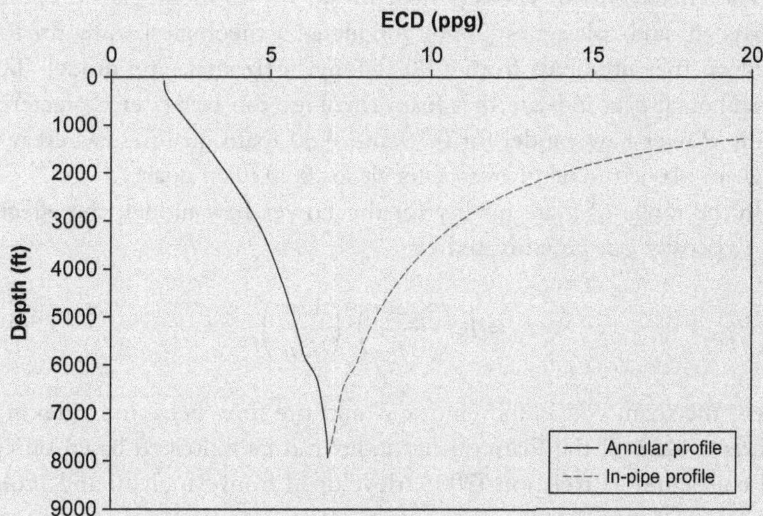

Figure 9.6 Calculated ECD profile along the flow path.

Foam Drilling Model

Guo and colleagues' foam drilling model (2003) is capable of simulating gas, water, oil, and solid 4-phase flows. The model takes the same form as Eq. (9.3). However, Eq. (9.11), derived for turbulent flow of aerated liquids, is not valid for foam due to different flow regimes and rheological models for foam. It can be shown that foams undergo laminar flows in most UBD conditions. The friction factor for a laminar flow is expressed as

$$f = \frac{64}{\text{Re}} \tag{9.15}$$

where the Reynolds number is expressed as

$$\text{Re} = \frac{\rho_f D_H v_f}{\mu_e} \tag{9.16}$$

where

ρ_f = foam density
D_H = hydraulic diameter of conduit
v_f = foam velocity
μ_e = effective foam viscosity

The effective foam viscosity depends on foam's rheological properties. Ozbayoglu and colleagues (2000) conducted a rheological study for foam based on measurements from a 90-ft-long horizontal pipe model. Their experimental data indicate that foam rheology can be better characterized by the Power Law model for 0.70 and 0.80 foam qualities, whereas the Bingham plastic model gives a better fit for 0.90 foam quality.

In the range of foam quality for the Power Law model, the effective foam viscosity can be estimated by

$$\mu_e = K\left(\frac{2n+1}{3n}\right)^n \left(\frac{12v_f}{D_H}\right)^{n-1} \tag{9.17}$$

where the foam consistency index K and the flow behavior index n for different values of the foam quality index can be estimated based on Guo and colleagues' correlation (2003) developed from Sanghani and Ikoku's experimental data (1982):

$$\begin{aligned} K = {} & -0.15626 + 56.147\Gamma - 312.77\Gamma^2 + 576.65\Gamma^3 + 63.960\Gamma^4 \\ & -960.46\Gamma^5 - 154.68\Gamma^6 + 1670.2\Gamma^7 - 937.88\Gamma^8 \end{aligned} \tag{9.18}$$

and

$$n = 0.095932 + 2.3654\Gamma - 10.467\Gamma^2 + 12.955\Gamma^3 + 14.467\Gamma^4 - 39.673\Gamma^5 + 20.625\Gamma^6 \quad (9.19)$$

where the in situ foam quality index Γ is defined as

$$\Gamma = \frac{\frac{2116}{P} Q_{go}}{\frac{2116}{P} Q_{go} + \frac{1}{7.48} Q_m + \frac{5.615}{60} Q_f} \quad (9.20)$$

or

$$\Gamma = \frac{GLR}{1 + GLR} \quad (9.21)$$

It is understood that the K-Γ relation and the n-Γ relation given by Eqs. (9.18) and (9.19), respectively, can be different for different types of foams (foams formed by different gases, liquids, and chemicals).

In the range of foam quality for the Bingham plastic model, the effective foam viscosity can be estimated by

$$\mu_e = \mu_p + \frac{g_c \tau_o D_H}{6 v_f} \quad (9.22)$$

where μ_p = plastic viscosity, and τ_o = yield point.

Illustrative Example 9.2

A well was cased to 5,280 ft with an 8-⅝-in, 28-lb/ft (8.017-in ID) casing. Starting from the kickoff point at 5,315 ft, the hole was drilled with mud using a 7⅞-in bit to build an inclination angle at a constant build rate of 4°/100 ft until the maximum inclination angle of 90 degrees was reached at a depth of 7,558 ft. Then the drilling fluid was shifted from mud to stable foam to drill to the TD of 8,048 ft while the inclination angle was maintained at 90 degrees. Additional data are given as follows. Calculate and plot the profiles of pressure, velocity, foam quality index, mixture density, and ECD along the flow path when the bit was at the total depth.

(Continued)

Illustrative Example 9.2 *(Continued)*

Drill string data

Length of heavy drill pipe below collar: 1,500 ft

Heavy drill pipe OD: 5 in

Heavy drill pipe ID: 3 in

Drill collar length: 480 ft

Drill collar OD: 6.25 in

Drill collar ID: 2.5 in

Drill pipe OD: 4.5 in

Drill pipe ID: 3.643 in

Material properties

Specific gravity of rock: 2.65 (water = 1)

Specific gravity of gas: 1 (air = 1)

Density of injected liquid: 8.4 ppg

Specific gravity of formation water: 1.05 (water = 1)

Specific gravity of formation oil: 0.85 (water = 1)

Pipe roughness: 0.0018 in

Borehole roughness: 0.1 in

Environment

Site elevation (above mean sea level): 50 ft

Ambient pressure: 14.7 psia

Ambient temperature: 60°F

Relative humidity: 0.1 fraction

Geothermal gradient: 0.01 F/ft

Friction tuning factor: 8

Operating conditions

Surface choke/flow line pressure: 14.7 psia

Rate of penetration: 120 ft/hour

Gas injection rate: 1,000 scfm

Liquid injection rate: 100 gpm

Formation water influx rate: 1 bbl/hour

Formation oil influx rate: 1 bbl/hour

Bit orifices: 20-1/32nd in
20-1/32nd in
20-1/32nd in

Solution

This problem can be solved with computer program *FoamDrill-09.xls*. The results are shown in Figures 9.7 through 9.11.

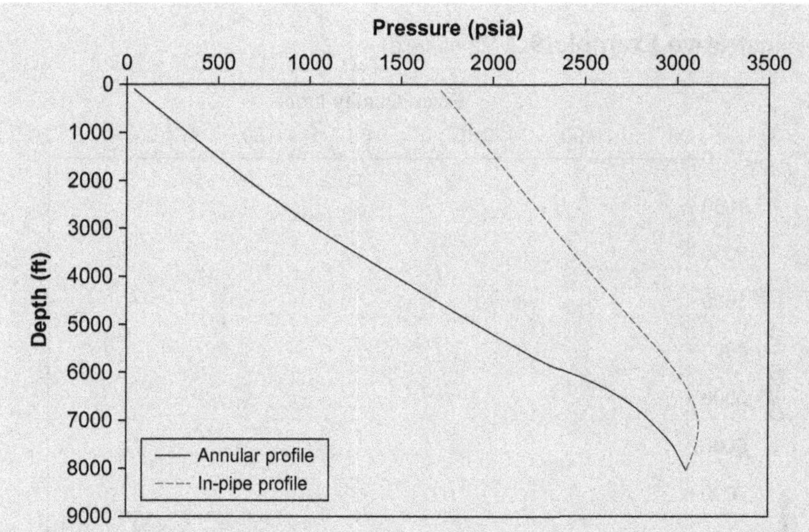

Figure 9.7 Calculated pressure profile along the flow path.

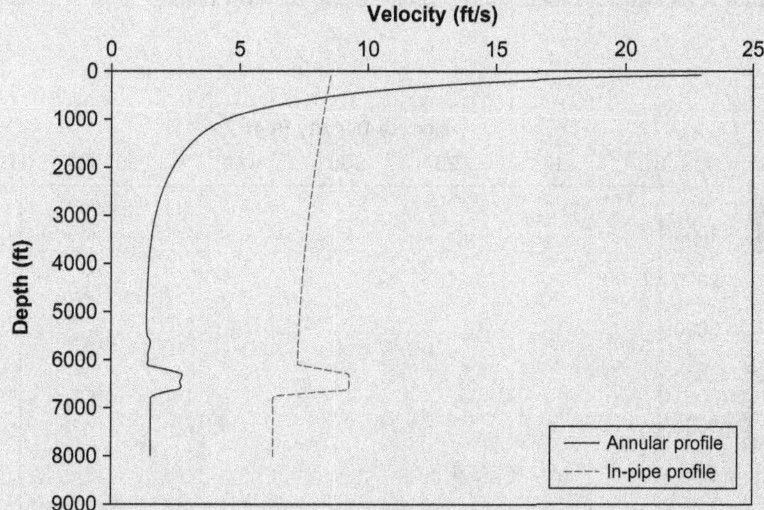

Figure 9.8 Calculated velocity profile along the flow path.

(Continued)

Illustrative Example 9.2 (*Continued*)

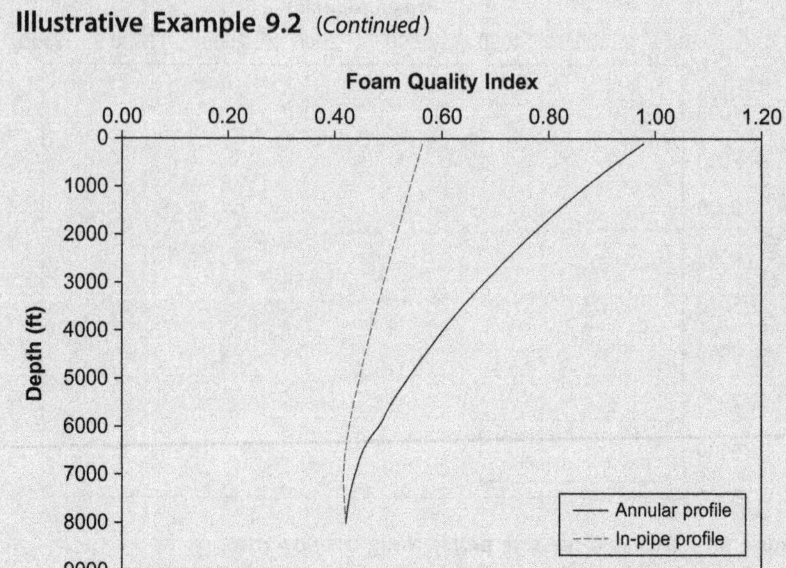

Figure 9.9 Calculated foam quality profile along the flow path.

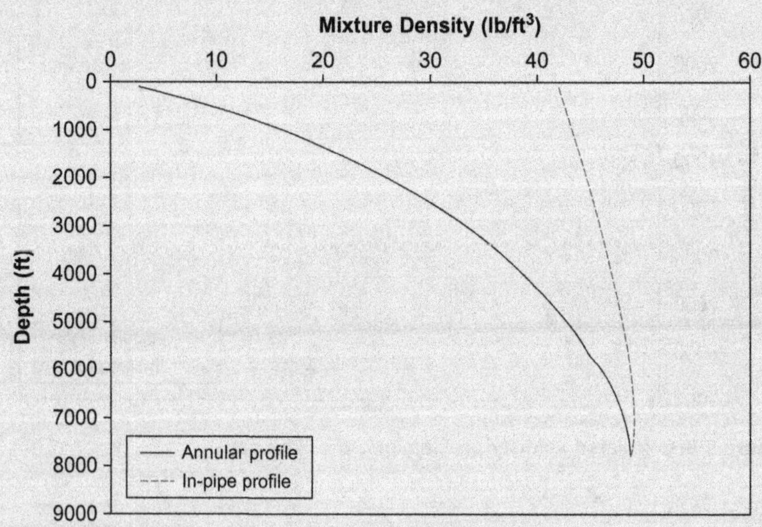

Figure 9.10 Calculated density profile along the flow path.

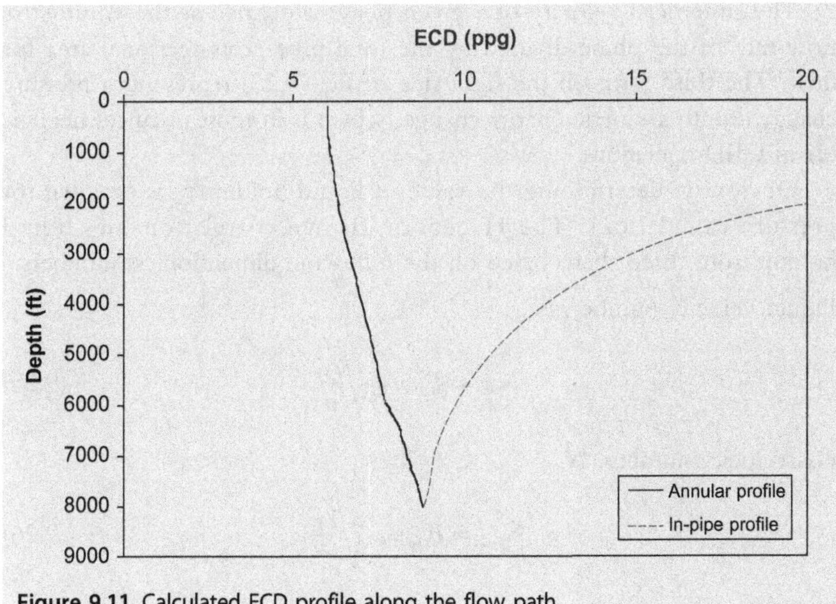

Figure 9.11 Calculated ECD profile along the flow path.

The Hagedorn–Brown Separated Flow Model

The Hagedorn–Brown (1965) correlation is widely used in petroleum production engineering for tubing performance calculations. It was proposed by Lyons and colleagues (2009) for aerated liquid drilling operations. The Hagedorn–Brown correlation takes the following form in U.S. field units:

$$144\frac{dp}{dz} = \gamma_{mix} + \frac{fW_t^2}{2.97\times 10^{11} D_H^5 \gamma_{mix}} + \gamma_{mix}\frac{\Delta(u_m^2)}{2g_c\Delta z} \tag{9.23}$$

where

W_t = total weight flow rate, lbf/day
γ_{mix} = in situ mixture specific weight, lbf/ft^3
u_m = mixture velocity, ft/s

and

$$u_m = u_{SL} + u_{SG} \tag{9.24}$$

where

u_{SL} = superficial velocity of liquid phase, ft/s
u_{SG} = superficial velocity of gas phase, ft/s

The superficial velocity of a given phase is defined as the volumetric flow rate of the phase divided by the total pipe cross-sectional area for flow. The third term on the right side of Eq. (9.23) represents a pressure change due to a kinetic energy change, which is in most instances negligible in UBD operations.

Obviously, determining the value of liquid holdup y_L is essential for pressure calculations. The Hagedorn–Brown correlation uses liquid holdup from three charts based on the following dimensionless numbers:

Liquid velocity number, $N_{\nu L}$:

$$N_{\nu L} = F_{\nu L} u_{SL} \sqrt[4]{\frac{\gamma_L}{\sigma}} \tag{9.25}$$

Gas velocity number, $N_{\nu G}$:

$$N_{\nu G} = F_{\nu G} u_{SG} \sqrt[4]{\frac{\gamma_L}{\sigma}} \tag{9.26}$$

Pipe diameter number, N_D:

$$N_D = F_D D_H \sqrt{\frac{\gamma_L}{\sigma}} \tag{9.27}$$

Liquid viscosity number, N_L:

$$N_L = F_L \mu_L \sqrt[4]{\frac{1}{\gamma_L \sigma^3}} \tag{9.28}$$

In U.S. field units, the unit conversion factors are

$$F_{\nu L} = 1.938$$
$$F_{\nu G} = 1.938$$
$$F_D = 120.872$$
$$F_L = 0.15726$$

In SI units, the unit conversion factors are

$$F_{\nu L} = 1.7964$$
$$F_{\nu G} = 1.7964$$
$$F_D = 31.664$$
$$F_L = 0.55646$$

The first chart in Hagedorn and Brown's (1965) work is used for determining the value of parameter (CN_L) based on N_L. Guo and colleagues

(2007) found that this chart can be replaced by the following correlation with good accuracy:

$$(CN_L) = 10^Y \tag{9.29}$$

where

$$Y = -2.69851 + 0.15841X_1 - 0.55100X_1^2 + 0.54785X_1^3 - 0.12195X_1^4 \tag{9.30}$$

and

$$X_1 = \log(N_L) + 3 \tag{9.31}$$

Once the value of parameter (CN_L) is determined, it is used for calculating the value of the group $\frac{N_{\nu L}p^{0.1}(CN_L)}{N_{\nu G}^{0.575}p_a^{0.1}N_D}$, where p is the absolute pressure at the location where the pressure gradient is to be calculated, and p_a is the atmospheric pressure. The value of this group is then used as an entry in the second chart in Hagedorn and Brown's model to determine parameter (y_L/ψ). Guo and colleagues (2007) found that the second chart can be represented by the following correlation with good accuracy:

$$\left(\frac{y_L}{\psi}\right) = -0.10307 + 0.61777[\log(X_2)+6] - 0.63295[\log(X_2)+6]^2 + 0.29598[\log(X_2)+6]^3 - 0.0401[\log(X_2)+6]^4 \tag{9.32}$$

where

$$X_2 = \frac{N_{\nu L}p^{0.1}(CN_L)}{N_{\nu G}^{0.575}p_a^{0.1}N_D} \tag{9.33}$$

According to Hagedorn and Brown, the value of parameter ψ can be determined from the third chart using a value of group $\frac{N_{\nu G}N_L^{0.38}}{N_D^{2.14}}$.

Guo and colleagues (2007) found that for $\frac{N_{\nu G}N_L^{0.38}}{N_D^{2.14}} > 0.01$ the third chart can be replaced by the following correlation with good accuracy:

$$\psi = 0.91163 - 4.82176X_3 + 1232.25X_3^2 - 22253.6X_3^3 + 116174.3X_3^4 \tag{9.34}$$

where

$$X_3 = \frac{N_{\nu G}N_L^{0.38}}{N_D^{2.14}} \tag{9.35}$$

However, $\psi = 1.0$ should be used for $\frac{N_{vG}N_L^{0.38}}{N_D^{2.14}} \leq 0.01$.

Finally, the liquid holdup can be calculated by

$$y_L = \psi\left(\frac{y_L}{\psi}\right) \tag{9.36}$$

The interfacial tension is a function of pressure and temperature. The following correlation is employed by Lyons and colleagues (2009):

$$\sigma = \sigma_{74} - \frac{(\sigma_{74} - \sigma_{280})(t - 74)}{206} \tag{9.37}$$

where the temperature t is in °F, or

$$\sigma = \sigma_{74} - \frac{(\sigma_{74} - \sigma_{280})(1.8t_C - 42)}{206} \tag{9.38}$$

where temperature t_C is in °C, and

$$\sigma_{74} = 75 - 1.108p^{0.349} \tag{9.39}$$

and

$$\sigma_{280} = 53 - 0.1048p^{0.637} \tag{9.40}$$

where pressure p is in psi, or

$$\sigma_{74} = 75 - 6.323{p_{MPa}}^{0.349} \tag{9.41}$$

and

$$\sigma_{280} = 53 - 2.517{p_{MPa}}^{0.637} \tag{9.42}$$

where pressure p_{MPa} is in MPa.

Illustrative Example 9.3

A well was cased to 5,552 ft with an 8⅝-in, 28-lb/ft (8.017-in ID) casing. Starting from the kickoff point at 5,575 ft, the hole was drilled with mud using a 7⅞-in bit to build an inclination angle at a constant build rate of 5°/100 ft until the maximum inclination angle of 90 degrees was reached at the depth of 7,358 ft. Then the drilling fluid was shifted from mud to stable foam to drill to the TD of 8,045 ft while the inclination angle was maintained at 90 degrees.

Additional data are given as follows. Calculate the pressure profile along the annulus.

Total measured depth: 8,045 ft
Hole roughness: 0.0018 in
Liquid injection rate: 191 gpm
Gas injection rate: 301 scf/min
Oil influx rate: 1 bbl/hour
Weight of injected liquid: 8.4 ppg
Gas specific gravity: 0.7
Oil gravity: 35°API
Liquid–gas interfacial tension: 60
Liquid viscosity: 2 cp
Oil viscosity: 1 cp
Surface temperature: 60°F
Flowing bottom hole temperature: 140°F
Surface choke pressure: 14.7 psia
Length of heavy drill pipe below collar: 425 ft
Heavy drill pipe OD: 5 in
Drill collar length: 500 ft
Drill collar OD: 5.75 in
Drill pipe OD: 4.5 in

Solution

This problem can be solved with computer program *Hagedorn-Brown Annular Pressure.xls*. The results are shown in Figure 9.12.

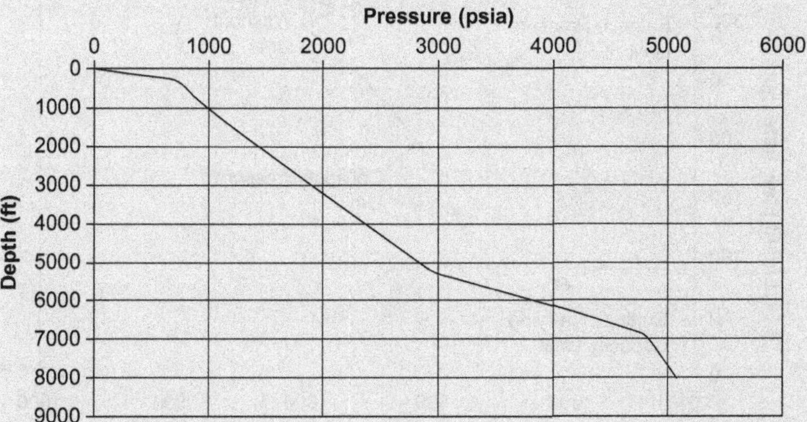

Figure 9.12 Calculated pressure profile in the annulus.

9.3 THE GAS–LIQUID FLOW RATE WINDOW

For a given hole geometry (hole and pipe sizes and depth) and fluid properties, the surface choke pressure, liquid flow rate, and gas injection rate are three major parameters that affect bottomhole pressures. The liquid flow rate and gas injection rate should be carefully designed to ensure underbalanced drilling and wellbore integrity. The gas–liquid flow rate window (GLRW) described in this section defines the margins of useable liquid and gas flow rates in underbalanced drilling.

The concept of GLRW was first defined by Guo and Ghalambor (2002). The combination of liquid flow rate and gas injection rate should be carefully designed so the flowing bottomhole pressure is less than the formation pore pressure under drilling conditions and the circulation-break bottomhole pressure is greater than the formation collapse pressure. Other considerations in designing liquid and gas flow rates include the cuttings carrying capacity of the fluid mixture and the wellbore washout. The former defines the lower bound of useable flow rate combinations, and the latter defines the upper limit of the useable flow rate combinations.

A typical GLRW is shown in Figure 9.13. The left boundary of the GLRW is determined by a locus of gas–liquid rate combinations that

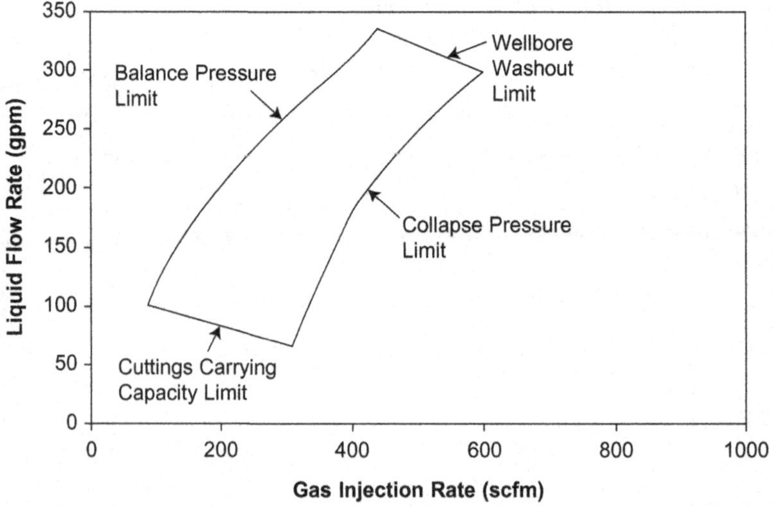

Figure 9.13 A typical gas–liquid flow rate window.

yield the same flowing bottomhole pressure being equal to the formation pore pressure minus a minimum UBD pressure differential. The right boundary of the GLRW is defined by a locus of gas–liquid rate combinations that yield the same circulation break bottomhole pressure which is equal to the formation collapse pressure at different gas injection rates. The lower boundary of the GLRW can be defined based on the cuttings carrying capacity of the fluid mixture. Different criteria for cuttings transport can be used, depending on the types of fluids.

A conservative criterion for aerated liquid is the minimum kinetic energy, which assumes that a minimum unit kinetic energy of 3 lbf-ft/ft^3 is required for drilling fluids to effectively carry drill cuttings up to the surface in normal drilling conditions. For calculation simplicity, it is safe to assume that the gas phase has no contribution to the carrying capacity of the mixture. This means that the minimum kinetic energy of an aerated fluid can be conservatively estimated based on the liquid flow rate. The upper boundary of the GLRW can be defined based on the wellbore washout constraint. Because no design method is available for this issue, a good practice is to look at calliper logs and use experience gained from past local drilling operations.

Determining the right and left boundaries of the GLRW requires knowledge of the UBD pressure differential and borehole stability against a hole collapse. These issues are discussed in the following sections.

9.3.1 The Underbalanced Drilling Pressure Differential

The UBD is mainly used for reducing formation damage during drilling. It is believed that the higher the pressure differential between the formation and the wellbore, the less the formation damage and thus the higher the well productivity. However, the pressure differential is also responsible for some drilling complications such as borehole collapse and excess formation fluid influx. Guo (2002) discussed the balance between formation damage and wellbore damage. Guo and Ghalambor (2006) presented a guideline for optimizing pressure differential in UBD and reported the following:

- The pressure differential should be high enough to counter the capillary pressure force responsible for liquid imbibition into the pay zone causing formation damage.
- The pressure differential should be low enough to ensure that the entire section of the open hole will not collapse during drilling.

- The pressure differential should be low enough to ensure that separators and storage facilities can handle the formation fluid influx rate and the total fluid volume.

For a water-wet reservoir, the capillary pressure at the sand face always causes the water in the drilling fluid to imbibe into the reservoir. This water imbibition increases water saturation in the reservoir and reduces the effective permeability to the hydrocarbon phase, which is referred to as filtration-induced formation damage. Figure 9.14 shows an imbibition capillary pressure curve for a typical sandstone. Initial water saturation exists in the sandstone at the moment when the rock is being drilled (point A). The capillary pressure at the initial condition is the highest, which is the potential to cause fast water imbibition into the rock. However, as the water saturation increases in the rock, the capillary pressure drops rapidly (point B). The water invasion at point B may not cause significant damage to the effective permeability to oil because the water has only occupied the narrow pore space ("corners") in the rock.

As imbibition continues, water will take over all the small pore space and begin to occupy the larger pore space (point C). Point C is a critical point because significant formation damage due to water invasion should occur beyond this point. The capillary pressure at point C is referred to the *critical* capillary pressure. It is this critical capillary pressure that should

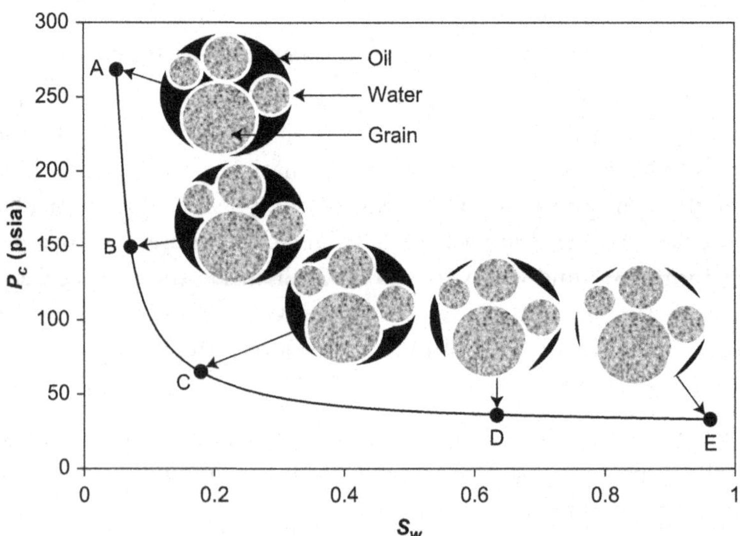

Figure 9.14 Capillary pressure curve for a sandstone sample in an air-brine system.

be balanced by the UBD pressure differential to prevent further water imbibition. If the underbalanced drilling pressure differential is less than the critical capillary pressure, water imbibition will continue, and significant formation damage is expected to occur (points D and E).

The existence of the critical capillary pressure can be proven on the basis of capillary parachor. Figure 9.15 shows a plot of parameter group $\left(\frac{S_{nw}}{P_c^2}\right)$ against S_w, where S_{nw} and S_w are nonwetting and wetting phase saturation, respectively, and P_c is imbibition capillary pressure. The peak value $\left(\frac{S_{nw}}{P_c^2}\right)_{max}$ was defined by Guo and colleagues (2004) as capillary parachor. The significance of capillary parachor is that it corresponds to a critical point where the nonwetting phase begins to lose its dominating state of flow. The authors observed the following:

1. Permeability (in md) of porous media is directly proportional to the Hg-air capillary parachor (in Atm^{-2}) with a proportionality factor of 0.054.
2. Permeability (in md) is proportional to the air-brine capillary parachor (in Atm^{-2}) squared, with a proportionality factor of 0.00007.

These observations imply that the capillary parachor is a parameter reflecting the effective pore size distribution of a porous medium for

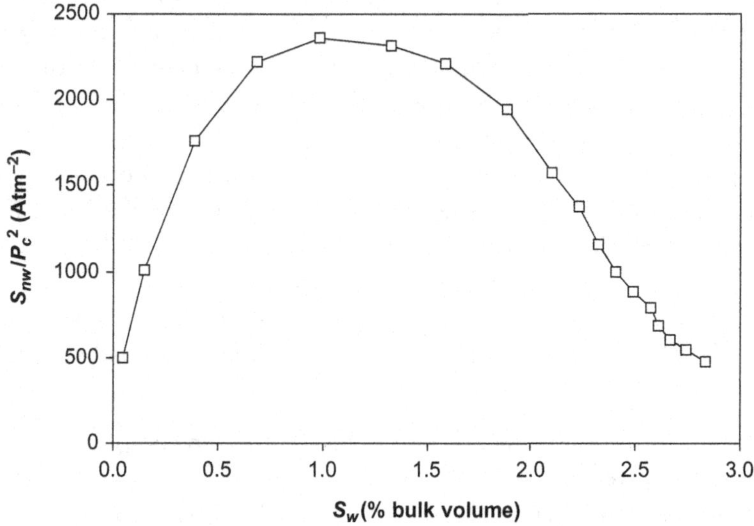

Figure 9.15 Plot of parameter $\frac{S_{nw}}{P_c^2}$ versus S_w for a sandstone sample in an air-brine system.

the dominant fluid. Thus, the point defining the capillary parachor is used to represent the critical point (point C) on the capillary pressure curve.

Owing to the parabolic nature of the capillary pressure curve, it is usually impossible to use a UBD pressure differential that is high enough to completely balance the capillary pressure at the initial water saturation. This is especially true for low permeability reservoirs, where the capillary pressure is normally very high at the initial water saturation. Fortunately, such a high-UBD pressure differential is not required because the capillary pressure drops rapidly with a small amount of water imbibition that should not cause significant formation damage. The UBD pressure differential should be just high enough to balance the critical capillary pressure defined at the capillary parachor. It is possible to keep the formation damage minimal using the critical pressure differential that is equal to the critical capillary pressure.

9.3.2 Collapse Pressure Analysis

A number of methods are available in the literature for wellbore collapse analyses. These methods include those used by Bradley (1979) and Aadnoy and Chenevert (1987). McLean and Addis (1990a) presented a review of these methods. The major difference among these methods is that they use different rock failure criteria such as Mohr-Coulomb, von Mises, Outer Drucker-Prager, Middle Drucker-Prager, and Inner Middle Drucker-Prager. McLean and Addis (1990b) presented their studies on the effect of rock strength criteria on mud weight recommendations. While the Mohr-Coulomb criterion suggests that an inclined borehole collapses first at the lateral sides of the hole, other criteria imply that an inclined borehole collapses first at the top side of the hole. More evidence from horizontal well drilling supports the Mohr-Coulomb criterion.

Borehole collapse analysis involves comparing failure stress to failure strength. The following data are required to complete a borehole collapse analysis:

- In situ formation stresses (both magnitude and direction)
- Pore pressure
- Rock elastic properties (elastic modulus and Poisson's ratio)
- Rock failure parameters (shear strength, tensile strength, and friction angle)

The in situ formation stresses are expressed as

$$\{\sigma\} = \begin{bmatrix} \sigma_H & 0 & 0 \\ 0 & \sigma_h & 0 \\ 0 & 0 & \sigma_V \end{bmatrix} \tag{9.43}$$

in which σ_H = the maximum horizontal stress; σ_h = the minimum horizontal stress; and σ_V = the vertical stress. The vertical stress can be estimated using density log numerically:

$$\sigma_V = \frac{1}{144}\int_0^D g\rho dz \approx \frac{g}{144}\sum_1^n \rho_i \Delta z_i \tag{9.44}$$

The minimum horizontal stress can be determined from mini-frac testing. The procedure for conducting a mini-frac test in a vertical borehole is shown in Figure 9.16. Hydraulic pressure is applied to the borehole by injecting water through the DST string until a mini-fracture is created. If a near-vertical fracture is created, the fracture is expected to break down the bottomhole. Oriented coring of the bottomhole should give the information about the fracture orientation. The fracture should be oriented in the direction of the maximum horizontal stress or opened against the minimum horizontal stress. The magnitude of the minimum horizontal stress can be determined from the pressure data recorded during the mini-frac testing. Figure 9.17 shows a typical shape of the pressure curves recorded in mini-frac tests. The magnitude of the initial shut-in pressure (ISIP) is taken as an approximation of the minimum horizontal stress.

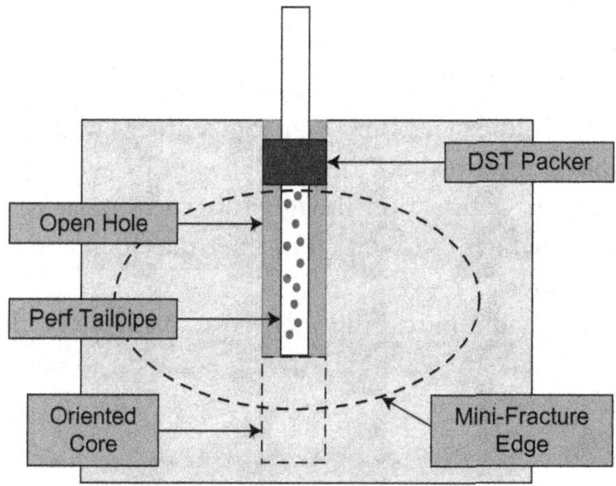

Figure 9.16 DST string for mini-frac testing.

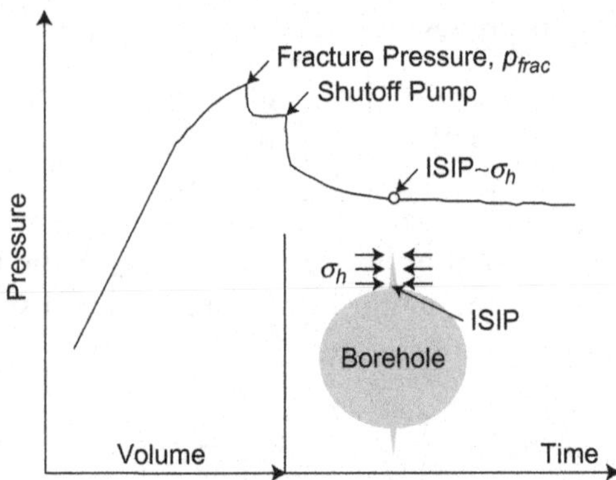

Figure 9.17 A typical shape of pressure curves recorded in mini-frac tests.

If the created fracture is not near-vertical, the fracture is not expected to break down the bottomhole. The oriented coring of the bottomhole will not give information about the fracture orientation. In such situations, the nonbroken cores can be taken to laboratories for analysis of the orientation of the maximum horizontal stress. Rock elastic properties (elastic modulus and Poisson's ratio) and rock failure parameters (shear strength, tensile strength, and friction angle) are also determined from laboratory tests.

Experimental methods used for estimating the orientation of the maximum horizontal stress include the following:

- Anelastic strain recovery (ASR) method
- Differential strain analysis (DSA) method
- Ultrasonic velocity analysis (UVA) method
- Paleomagnetics method
- Core eccentricity method

Oriented 4-arm caliper and dipmeter logs with MWD can be used to identify the borehole breakout that is believed to align with the minimum horizontal stress direction. Imaging logs can identify fractures and breakout/failures, which can be used to infer the orientation of the minimum horizontal stress.

To date, no reliable measuring method is available to determine the maximum horizontal stress. The following equation can be used to estimate its value:

$$\sigma_H \approx 3\sigma_h + \sigma_Y - p_{frac} - p_{pore} \tag{9.45}$$

where σ_Y is the yield strength of rock.

Borehole collapse analyses start from calculating stresses on the borehole wall both without and with the borehole. Consider a situation where a borehole section is to be drilled with an inclination angle α in the direction of β degrees from the maximum horizontal stress, as illustrated in Figure 9.18. For any point at the imagined borehole wall at a loop angle θ from the highest point, the stresses before drilling the hole are expressed as

$$\{\hat{\sigma}\} = \begin{bmatrix} \sigma_x & \sigma_{xy} & \sigma_{xz} \\ \sigma_{xy} & \sigma_y & \sigma_{yz} \\ \sigma_{xz} & \sigma_{yz} & \sigma_z \end{bmatrix} \tag{9.46}$$

The stress components in Eq. (9.46) can be determined from the following transformation:

$$\{\hat{\sigma}\} = [Q]\{\sigma\}[Q]^T \tag{9.47}$$

where

$$[Q] = \begin{bmatrix} \cos\beta\cos\alpha & \sin\beta\cos\alpha & \sin\alpha \\ -\sin\beta & \cos\beta & 0 \\ -\cos\beta\sin\alpha & -\sin\beta\sin\alpha & \cos\alpha \end{bmatrix} \tag{9.48}$$

and

$$[Q]^T = \begin{bmatrix} c\cos\beta\cos\alpha & -\sin\beta & -\cos\beta\sin\alpha \\ \sin\beta\cos\alpha & \cos\beta & -\sin\beta\sin\alpha \\ \sin\alpha & 0 & \cos\alpha \end{bmatrix} \tag{9.49}$$

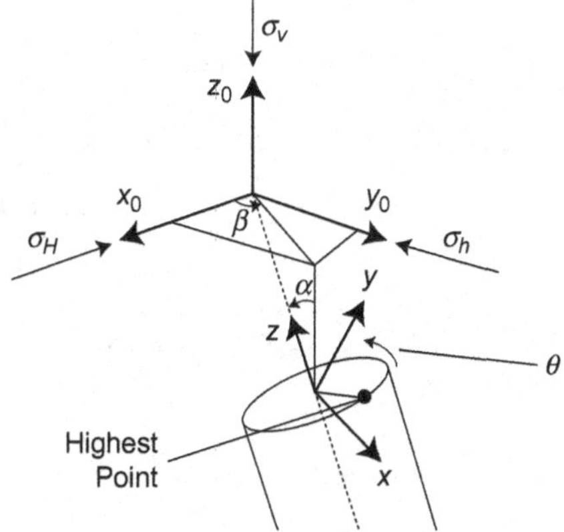

Figure 9.18 A borehole section is to be drilled with an inclination angle α in the direction of β degrees from the maximum horizontal stress.

The stresses at the same point during drilling of the hole are expressed as

$$\begin{cases} \sigma_r = p_{bh} \\ \sigma_\theta = \sigma_x + \sigma_y - 2(\sigma_x - \sigma_y)\cos(2\theta) - 4\sigma_{xy}\sin(2\theta) - p_{bh} \\ \sigma_a = \sigma_V - \nu[2(\sigma_x - \sigma_y)\cos(2\theta) + 4\sigma_{xy}\sin(2\theta)] \\ \sigma_{\theta a} = 2[\sigma_{yz}\cos(\theta) - \sigma_{xz}\sin(\theta)] \end{cases} \quad (9.50)$$

where σ_r, σ_θ, and σ_a are stresses in the radial, tangential (loop), and axial directions, respectively. It can be shown that the loop stress σ_θ reaches its maximum at $\theta = 90°$ and $\theta = 270°$. The principal stresses at the point of concern are expressed as

$$\sigma_1 = \frac{\sigma_\theta + \sigma_a}{2} + \sqrt{\left(\frac{\sigma_\theta - \sigma_a}{2}\right)^2 + \sigma_{a\theta}^2} \quad (9.51)$$

$$\sigma_2 = \frac{\sigma_\theta + \sigma_a}{2} - \sqrt{\left(\frac{\sigma_\theta - \sigma_a}{2}\right)^2 + \sigma_{a\theta}^2} \quad (9.52)$$

and

$$\sigma_3 = \sigma_r \quad (9.53)$$

The maximum and minimum stresses among the three principal stresses are expressed as

$$\sigma_{max} = \max(\sigma_1, \sigma_2, \sigma_3) \quad (9.54)$$

and

$$\sigma_{min} = \min(\sigma_1, \sigma_2, \sigma_3) \quad (9.55)$$

The Mohr-Coulomb failure criterion can be written as

$$\sigma'_{max} \le 2S_o \tan\left(\frac{\pi + 2\varphi}{4}\right) + \sigma'_{min} \tan^2\left(\frac{\pi + 2\varphi}{4}\right) \quad (9.56)$$

in which S_o is the cohesive strength and ϕ is the friction angle. The effective stress is expressed as

$$\sigma' = \sigma - \alpha_B p_{pore} \quad (9.57)$$

where α_B is the poroelastic constant taking a value between rock porosity and 1, averaging at 0.72.

Illustrative Example 9.4

A well is to be drilled through a pay zone at a true vertical depth of 4,505 ft with a pore pressure of 1,802 psia. Additional data are given as follows. Aerated liquid is to be used to achieve an underbalanced pressure differential of 300 psi, or borehole pressure 1,502 psia. If the wellbore is drilled in the direction of the minimum horizontal stress, what is the maximum allowable inclination angle in the pay zone in order to avoid borehole collapse?

In situ stresses

Vertical in situ stress, σ_V: 4,908 psi
Maximum horizontal stress, σ_H: 2,850 psi
Minimum horizontal stress, σ_h: 2,650 psi

Rock properties

Cohesive strength, S_o: 1,402 psi
Friction angle, ϕ: 39 Deg.
Poisson's ratio, v: 0.28
Poroelastic constant, a_p: 0.72

Solution

This problem can be solved with computer program *Mohr-Coulomb Collapse Gradient.xls*. The collapse envelope is plotted in Figure 9.19. The borehole pressure of 1,502 psia is equivalent to a fluid weight of 6.41 ppg. The figure shows that the 6.41-ppg fluid weight will cause the collapse of boreholes with inclination angles of higher than 45°.

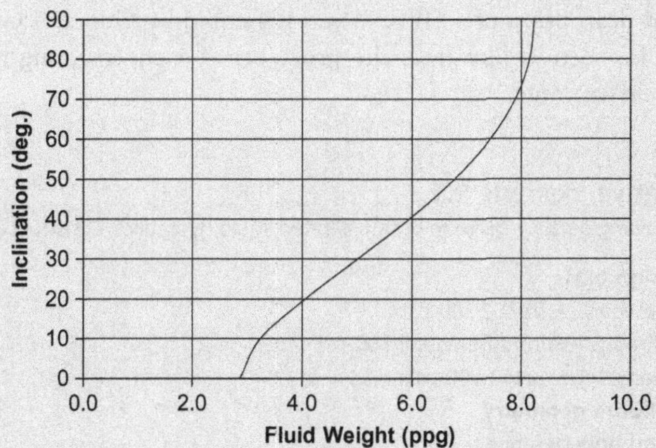

Figure 9.19 Calculated collapse envelope.

9.3.3 Predicting Fluid Influx

Reservoir fluid influx is another factor that limits the UBD pressure differential. Several methods have been used for predicting fluid influx in underbalanced drilling. Garham and Culen (2004) analyzed gas influx from a gas reservoir to the borehole during UBD. They presented an approach to predicting expected gas production rates, sizing equipment, and managing bottomhole pressure to aid in borehole stability and formation damage control. Single-phase gas flow was assumed in the borehole. Rommetveit and colleagues (2004) considered multiphase flow in the borehole in their dynamic model for predicting formation oil influx during UBD.

Mykytiw and colleagues (2004) used a multiphase flow simulator for UBD applications design with a steady flow reservoir model. Friedel and Voigt (2004) employed a numerical reservoir simulator to investigate gas inflow during underbalanced drilling and the impact of UBD on longtime well productivity considering the non-Darcy flow effect. Haghshenas (2005) analyzed the effects of drilling parameters and reservoir properties on the formation oil influx rate and total influx volume in UBD. He used radial and spherical transient flow models to estimate fluid influx during drilling. Guo and Shi (2007) and Guo and colleagues (2008) presented mathematical models for predicting influx rate and volume for planning UBD horizontal wells. These models can be used for adjusting UBD pressure differentials to fit the capacities of separators and fluid storage tanks.

9.3.4 Constructing a Gas–Liquid Flow Rate Window

A detailed description of a GLRW was presented by Guo and Ghalambor (2002). This section illustrates the procedure for constructing a GLRW using a field example.

Illustrative Example 9.5

The following data are given to design aerated liquid hydraulics. Construct a GLRW.

Design basis
Reservoir pressure: 2,300 psia
Desired pressure differential: 300 psi
Collapse pressure: 1,500 psia
Wellbore geometry
Cased hole depth: 5,000 ft
Casing ID: 8.125 in
Open hole diameter: 8 in
Vertical depth: 5,000 ft

Measured depth: 5,000 ft
Drill pipe OD: 4.5 in
Material properties
Specific gravity of solid: 2.7 (water = 1)
Liquid weight: 8.5 ppg
Specific gravity of formation fluid: 0.8 (water = 1)
Specific gravity of gas: 1 (air = 1)
Pipe roughness: 0.0018 in
Borehole roughness: 0.1 in
Productivity index: 3.333 bbl/d/psi
Environment
Ambient pressure: 14.7 psia
Ambient temperature: 70°F
Geothermal gradient: 0.01 F/foot
Operation conditions
Bit diameter: 7.875 in
Rotary speed: 50 rpm
Penetration rate: 60 ft/hr
Choke pressure: 40 psia

Solution

Circulation-breaking bottomhole pressures at various liquid flow rates and gas injection rates are calculated using Eq. (9.3) with the friction factor set to zero. The results are plotted in Figure 9.20. A horizontal line is drawn in the plot

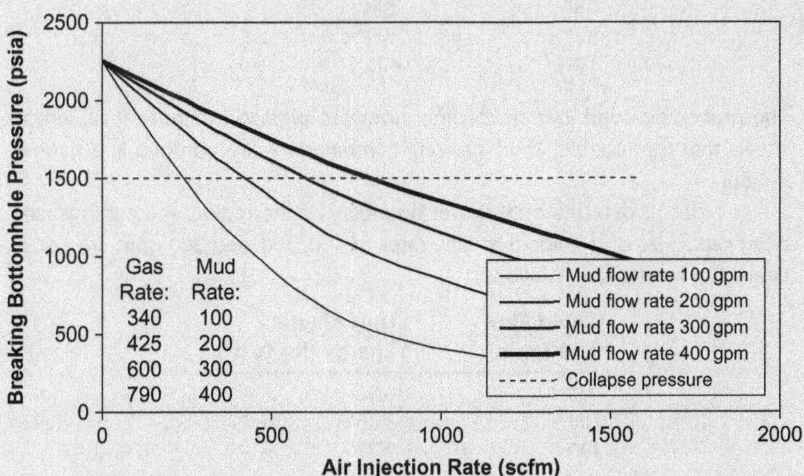

Figure 9.20 Calculated bottomhole pressure curves for circulation break conditions with identified flow rate combinations to yield borehole collapse: 7⅞-in bit, 4½-in pipe, 5,000-ft depth, 60-ft/hr ROP, 42-bbl/h oil influx.

(Continued)

Illustrative Example 9.5 *(Continued)*
at a circulation-breaking bottomhole pressure of 1,500 psia. This horizontal line intersects the pressure curves at the flowing points:

Gas Injection Rate (scfm)	Liquid Flow Rate (gpm)
340	100
425	200
600	300
790	400

These combinations of flow rates define the maximum allowable air injection rates at given mud rates so the borehole will not collapse. A curve (to be plotted later) going through these points defines the right boundary of a GLRW.

Flowing bottomhole pressures at various liquid flow rates and gas injection rates are calculated using Eq. (9.3), and the results are plotted in Figure 9.21. A horizontal line is drawn in the plot at a flowing bottomhole pressure of 2,000 psia. This horizontal line intersects the pressure curves at the flowing points:

Gas Injection Rate (scfm)	Liquid Flow Rate (gpm)
90	100
200	200
380	300
720	400

The preceding combinations of flow rates are plotted in Figure 9.22, which shows that the useable liquid–gas rate combinations are clamped in a narrow region.

In order to determine the lower boundary of the region, the cuttings carrying capacities of the liquid at flow rates of 100, 200, and 300 gpm are calculated. The results are as follows:

Liquid Flow Rate (gpm)	Unit Kinetic Energy (lbf-ft/ft^3)
100	0.86
200	3.46
300	7.77

The unit kinetic energy of 3 lbf-ft/ft^3 corresponds to a liquid flow rate of 187 gpm. These data points are plotted in Figure 9.23 to show their locations in the flow rate map.

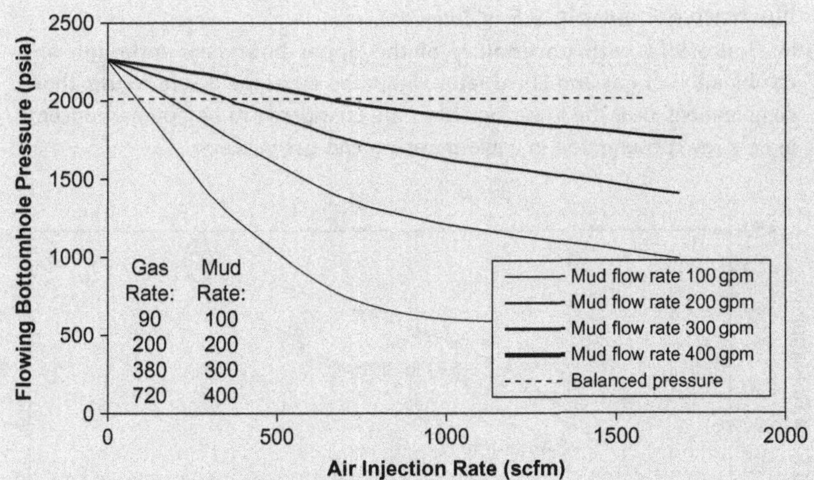

Figure 9.21 Calculated bottomhole pressure curves for normal drilling conditions with identified flow rate combinations to balance formation pore pressure: 7⅞-in bit, 4½-in pipe, 5,000-ft depth, 60-ft/hr ROP, 42-bbl/h oil influx.

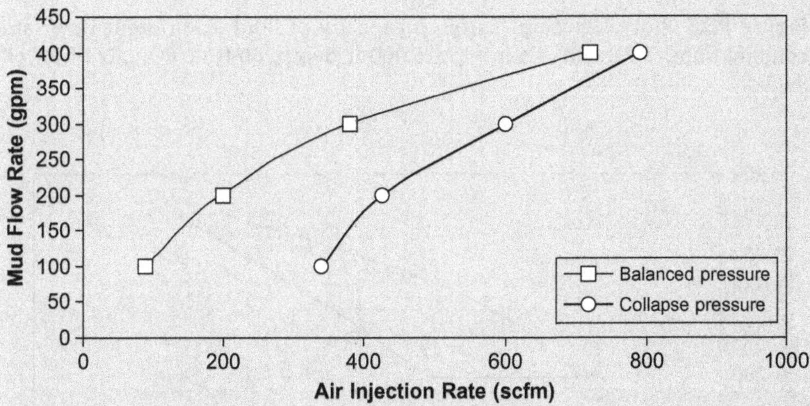

Figure 9.22 Calculated flow rate combination loci for balanced and collapse pressures: 7⅞-in bit, 4½-in pipe, 5,000-ft depth, 60-ft/hr ROP, 42-bbl/h oil influx.

Without knowing the geological details throughout the open hole section in the area and the wellbore washout experience, it is difficult to close the upper boundary of the flow rate envelope. The resultant GLRW is presented

(*Continued*)

Illustrative Example 9.5 (*Continued*)

in Figure 9.24 with uncertainty of the upper boundary. Although any combination of gas and liquid rates within the envelope is safe to use, those combinations near the lower boundary are considered to be optimal concerning energy consumption in liquid pumping and gas injection.

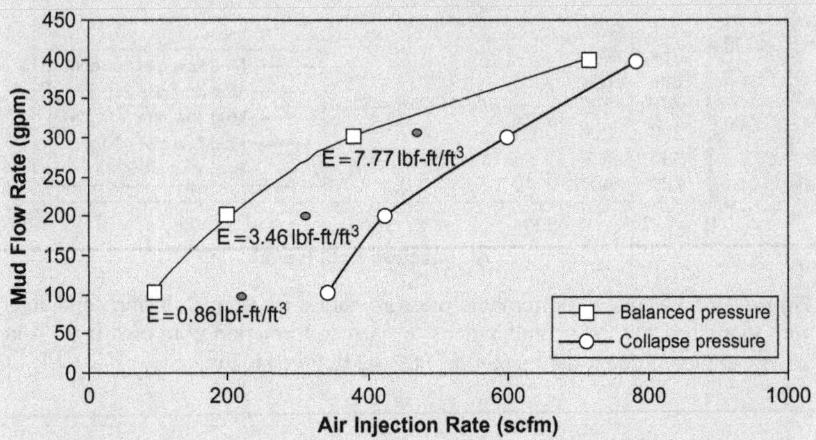

Figure 9.23 Plot of cuttings carrying capacity of fluid at different flow rate combinations: 7⅞-in bit, 4½-in pipe, 5,000-ft depth, 60-ft/hr ROP, 42-bbl/h oil influx.

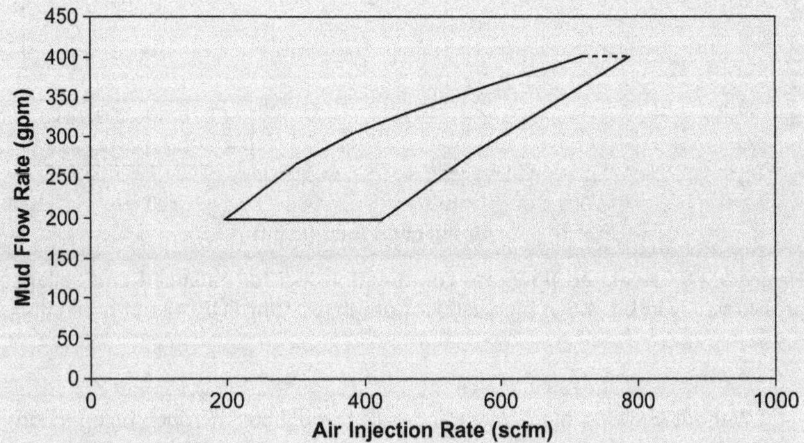

Figure 9.24 Calculated GLRW: 7⅞-in bit, 4½-in pipe, 5,000-ft depth, 60-ft/hr ROP, 42-bbl/h oil influx.

SUMMARY

This chapter described the principle of selecting the gas and liquid injection rates for underbalanced drilling operations. The concept of the gas–liquid rate window was explained and the procedure for constructing the GLRW was illustrated.

REFERENCES

Aadnoy, B.S., Chenevert, M.E., 1987. Stability of highly inclined boreholes. SPE Drill. Eng. (December), 364–374.

Bradley, W.B., 1979. Failure of include boreholes. J. Energy Resour. Technol., Trans. AIME 101, 232–239.

Caetano, E.F., Shoham, O., Brill, J.P., 1992. Upward vertical 2-phase flow through an annulus, part I: Single phase friction factor, Taylor bubble velocity, and flow pattern prediction. J. Energy Resour. Technol. 114, 1–13.

Friedel, T., Voigt, H., 2004. Numerical simulation of the gas inflow during underbalanced drilling (UBD) and investigation of the impact of UBD on longtime well productivity. In: Paper SPE/IADC 91558, Presented at the DPE/IADC Underbalanced Technology Conference and Exhibition, October 11–12, Houston.

Garham, R.A., Culen, M.S., 2004. Methodology for manipulation of wellhead pressure control for the purpose of recovering gas to process in underbalanced drilling applications. Paper SPE/IADC 91220, Presented at the DPE/IADC Underbalanced Technology Conference and Exhibition, October 11–12, Houston.

Govier, G.W., Aziz, K., 1977. The Flow of Complex Mixtures in Pipes. Robert E. Drieger Publishing Co.

Guo, B., 2002. Balance between formation damage and wellbore damage: What is the controlling factor in UBD operations? Paper SPE 73735, Proceedings of the SPE International Symposium and Exhibition on Formation Damage Control, February 20–21, Lafayette, LA.

Guo, B., Feng, Y., Ghalambor, A., 2008. Prediction of influx rate and volume for planning UBD horizontal wells to reduce formation damage. Paper SPE 111346, Proceedings of the SPE International Symposium and Exhibition on Formation Damage Control, February 13–15, Lafayette, LA.

Guo, B., Ghalambor, A., 2002. An innovation in designing underbalanced drilling flow rates: a gas-liquid rate window (GLRW) approach. Paper SPE 77237, Proceedings of the IADC/SPE Asia Pacific Drilling Technology, September 9–11, Jakarta.

Guo, B., Ghalambor, A., 2006. A guideline to optimizing pressure differential in underbalanced drilling for reducing formation damage. Paper SPE 98083, Proceedings of the SPE International Symposium and Exhibition on Formation Damage Control, February 15–17, Lafayette, LA.

Guo, B., Ghalambor, A., Duan, S., 2004. Correlation between sandstone permeability and capillary pressure curves. J. Pet. Sci. Eng. (August), 21.

Guo, B., Hareland, G., Rajtar, J., 1996. Computer simulation predicts unfavorable mud rate and optimum air injection rate for aerated mud drilling. In: SPEDC, pp. 61–66.

Guo, B., Lyons, W.C., Ghalambor, A., 2007. Petroleum Production Engineering. Elsevier.

Guo, B., Shi, X., 2007. A rigorous reservoir-wellbore cross-flow equation for predicting liquid inflow rate during underbalanced horizontal drilling. Paper SPE 108363, Proceedings of the SPE Asia Pacific Oil and Gas Conference and Exhibition, October 30–November 1, Jakarta.

Guo, B., Sun, K., Ghalambor, A., 2003. A closed form hydraulics equation for predicting bottom-hole pressure in UBD with foam. Paper SPE 81640, Proceedings of IADC/SPE Underbalanced Technology Conference and Exhibition, March 25–26, Houston.

Guo, B., Sun, K., Ghalambor, A., Xu, C., 2004. A closed form hydraulics equation for aerated mud drilling in inclined wells. SPE Drill. Completion, J. (June).

Hagedorn, A.R., Brown, K.E., 1965. Experimental study of pressure gradients occurring during continuous two-phase flow in small diameter vertical conduits. J. Pet. Technol. (April), 476–484.

Haghshenas, A., 2005. Influx volume calculation during underbalanced drilling. M.S. Thesis, University of Louisiana at Lafayette.

Lage, A.C.V.M., Time, R.W., 2000. Mechanistic model for upward two-phase flow in annuli. Paper SPE 63127, presented at the SPE Annual Technical Conference and Exhibition, October 1–4, Dallas.

Lage, A.C.V.M., Fjelde, K.K., Time, R.W., 2000. Underbalanced drilling dynamics: two-phase flow modeling and experiments. Paper SPE 62743, presented at the IADC/SPE Asia Pacific Drilling Technology Conference, September 11–13, Kuala Lumpur.

Lyons, W.C., Guo, B., Graham, R.L., Hawley, G.D., 2009. Air and Gas Drilling Manual, third ed. Elsevier.

Lyons, W.C., Guo, B., Seidel, F.A., 2001. Air and Gas Drilling Manual, second ed. McGraw-Hill.

McLean, M.R., Addis, M.A., 1990. Wellbore stability analysis: a review of current methods of analysis and their field applications. Paper IADC/SPE 19941, presented at the IADC/SPE Drilling Conference, February 27–March 2, Houston.

McLean, M.R., Addis, M.A., 1990. Wellbore stability analysis: the effect of strength criteria on mud weight recommendations. Paper SPE 20405, presented at the SPE Annual Technical Conference and Exhibition, September 23–26, New Orleans.

Mykytiw, C., Syryanarayana, P.V., Brand, P.R., 2004. Practical use of a multiphase flow simulator for underbalanced drilling applications design—the tricks of the trade. Paper SPE/IADC 91598, presented at the DPE/IADC Underbalanced Technology Conference and Exhibition, October 11–12, Houston.

Nakagawa, E.Y., Silva Jr., V., Boas, M.D.V., Silva, P.R.C., Shayegi, S., 1999. Comparison of aerated fluids/foam drilling hydraulics simulators against field data. Paper SPE 54319, presented at the SPE Asia Pacific Oil and Gas Conference and Exhibition, April 20–22, Jakarta.

Ozbayoglu, M.E., Kuru, E., Miska, S., Takach, N., 2000. A comparative study of hydraulic models for foam drilling. SPE Paper 65489, Proceedings of the SPE/PS CIM International Conference on Horizontal Well Technology, November 6–8, Calgary, AB.

Rommetveit, R., Fjelde, K.K., Froyen, J., Bjorkevoll, K.S., Boyce, G., Olsen, E., 2004. Use of dynamic modeling in preparation for the Gullfaks C-5A well. Paper SPE/IADC 91243, presented at the DPE/IADC Underbalanced Technology Conference and Exhibition, October 11–12, Houston.

Sanghani, V., 1982. Rheology of foam and its implications in drilling and cleanout operations. M.S. Thesis, University of Tulsa, Tulsa.

Sunthankar, A.A., Kuru, E., Miska, S., 2001. New developments in aerated mud hydraulics for drilling in inclined wells. Paper SPE 67189, presented at the SPE Production and Operations Symposium, March 24–27, Oklahoma City.

PROBLEMS

9.1 Solve the problem in Illustrative Example 9.1 with choke pressure 100 psia.

9.2 Solve the problem in Illustrative Example 9.2 with choke pressure 100 psia.

9.3 Solve the problem in Illustrative Example 9.3 with choke pressure 100 psia.

9.4 Assuming a rock porosity value of 10%, determine the critical capillary pressure based on the capillary curve shown in Figure 9.14.

9.5 Solve the problem in Illustrative Example 9.4 with cohesive strength 1,200 psi, friction angle 32°, and Poisson's ratio 0.22.

CHAPTER TEN

Underbalanced Drilling Operations

Contents

10.1 INTRODUCTION

Underbalanced drilling is commonly carried out using aerated liquids and foams. Since the operations are mostly for improving hydrocarbon recovery from depleted reservoirs, well control is usually not a major concern as long as the required types of well control equipment are installed. When drilling with aerated liquids, pressure stability and directional drilling problems may occur. Foam stability and foam disposal are important issues in foam drilling. This chapter provides some basic knowledge of unbalanced drilling (UBD) while focusing on minimizing drilling complications.

10.2 AERATED LIQUID DRILLING

Aerated liquid drilling is also referred to as aerated mud drilling or gasified drilling. Gas (air or nitrogen) is mixed with liquid (water, mud, or oil) in the borehole to reduce the bottomhole pressure for achieving underbalanced conditions. The level of bottomhole pressure and its stability mainly depend on the aeration method and the gas injection rate.

10.2.1 Aeration Methods

Several gas injection methods have been used in the industry to solve different types of problems:

- Drill string injection
- Parasite string injection
- Parasite casing injection
- Through completion injection

The last three methods are annulus injection methods. The drill string gas injection method is shown in Figure 10.1. Gas and liquid are both injected into the standpipe, and they mix while flowing down the drill string. The full annulus is occupied by aerated fluid. The advantages of this method include its simplicity and low cost. The major disadvantage of this method is the borehole washout in the openhole section because of the high velocity of liquid and gas mixture around the drill collar.

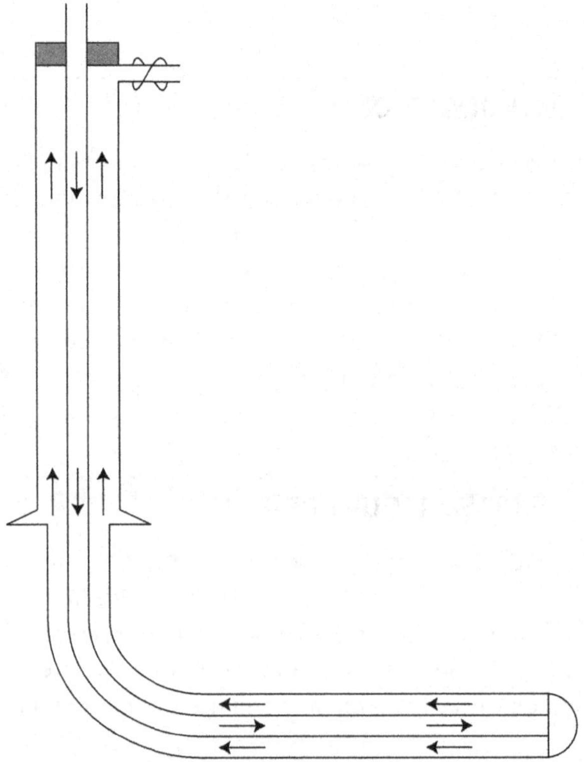

Figure 10.1 Drill string gas injection method.

To ease this problem, side-jet subs have been used to divert some of the gas flow in the cased hole section (Figure 10.2). Another disadvantage of the drill string injection method is that it prevents the use of conventional measurement while drilling (MWD) tools for directional drilling. This is because the compressibility of the aerated fluid in the drill string masks the mechanical telemetry signal from the MWD tool. Electric-magnetic measurement while drilling tools, or EMWD, are required with the drill string injection method. To solve this problem and to increase pressure stability, annular injection methods were developed.

The first annulus injection method developed was the parasite string injection method shown in Figure 10.3. Gas is injected into the cased annulus through a small tubing string attached to the casing string. The gas injection sub connecting the tubing string to the casing is shown in Figure 10.4. With this gas injection method, only the cased hole section of the annulus is aerated. It allows for the use of conventional MWD in directional drilling

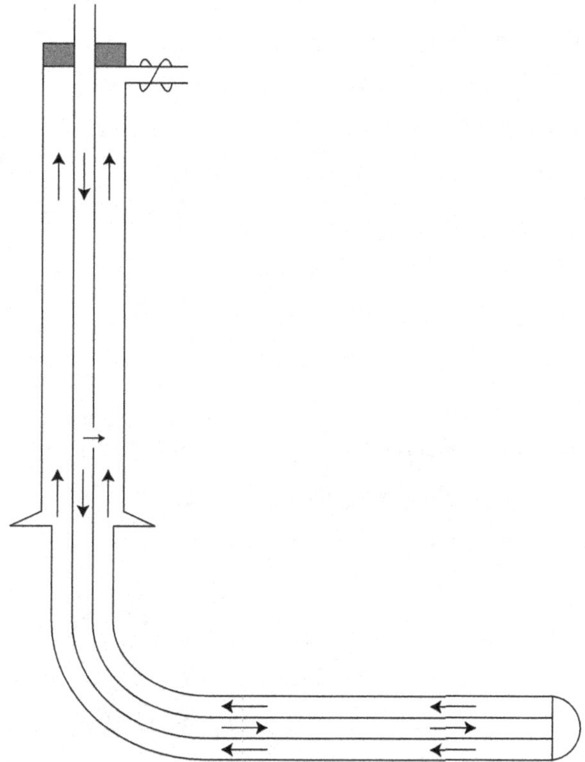

Figure 10.2 Drill string gas injection with a side-jet sub.

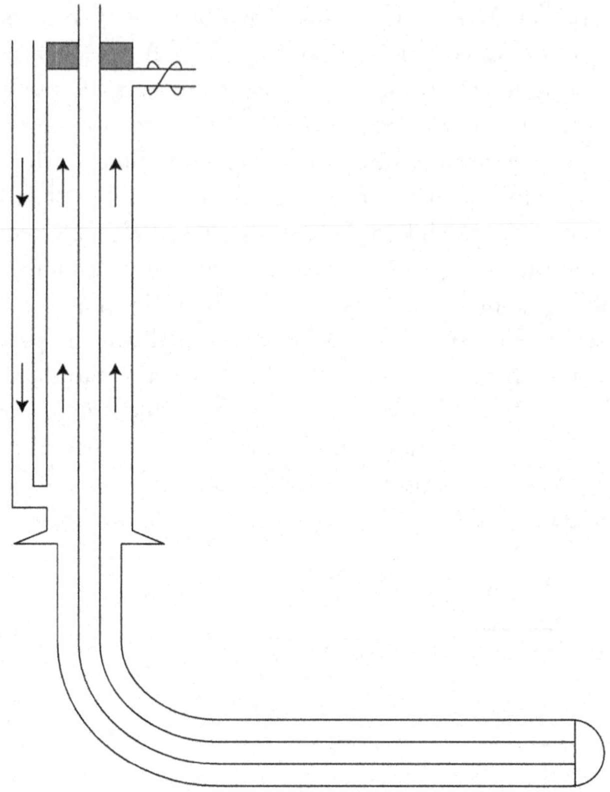

Figure 10.3 Parasite string gas injection method.

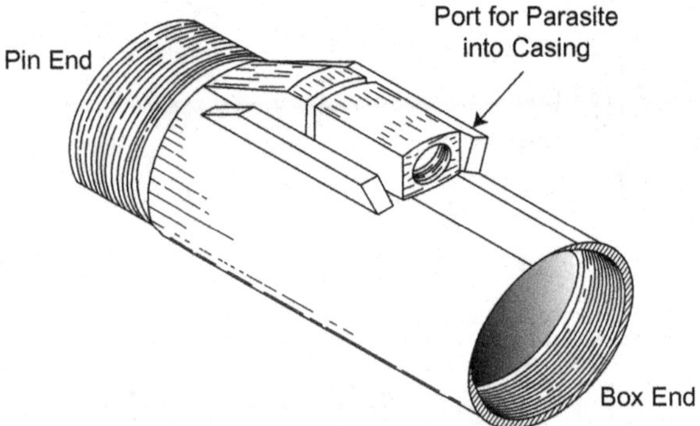

Figure 10.4 Gas injection sub used in parasite string gas injection.

and better pressure stability control at the bottomhole. However, running in casing with a parasite tubing string is difficult.

The second annulus injection method is the parasite casing injection method shown in Figure 10.5. The outer casing can be set and cemented beyond the vertical section. The inner casing is set but is not cemented. Gas is injected into the drill string–inner casing annulus through the inner–outer casing annulus. The inner-casing string is usually retrieved after drilling. With this gas injection method, only a portion of the cased hole section of the annulus is aerated. It again allows for the use of conventional MWD and better pressure stability control at the bottomhole. However, complications can occur due to the nonconstrained, noncentralized inner casing at the bottom.

The third annulus injection method is the through completion injection method shown in Figure 10.6. The outer casing can be set and

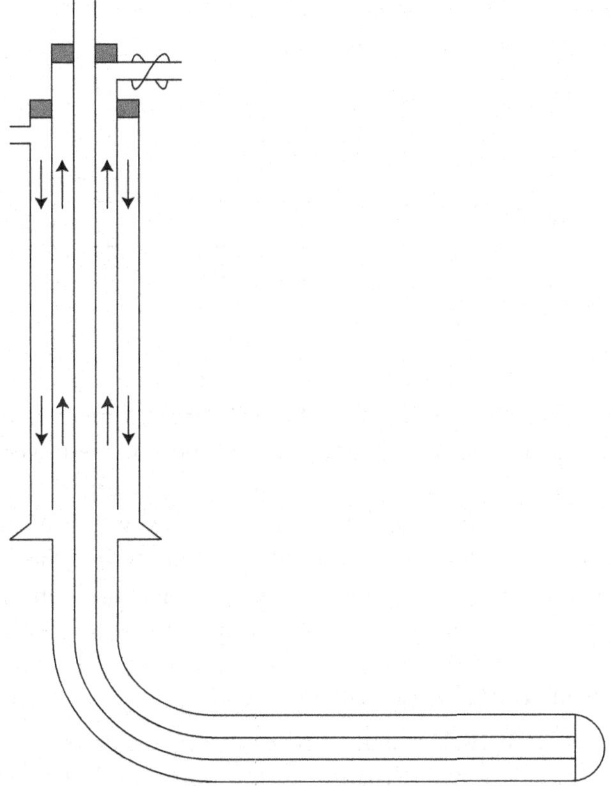

Figure 10.5 Parasite casing gas injection method.

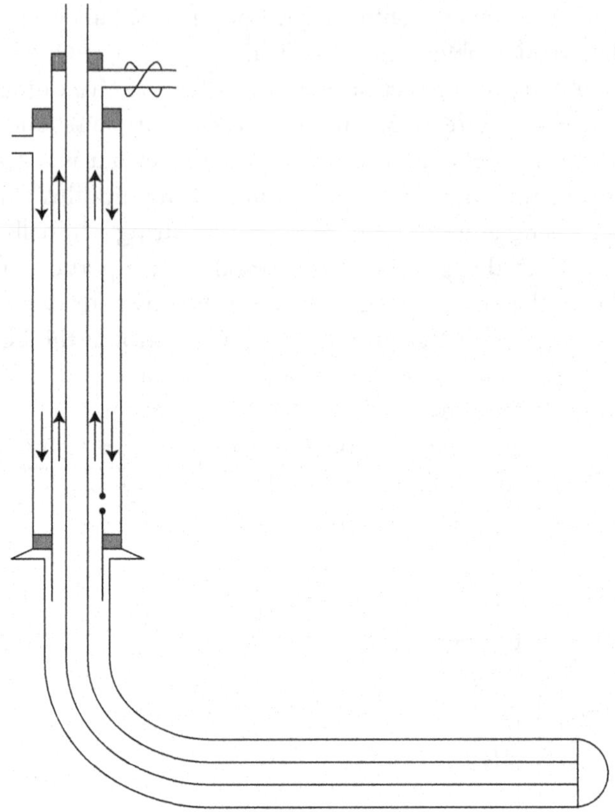

Figure 10.6 Through completion gas injection method.

cemented at any depth. The inner casing is set with a packer inside the outer casing. Gas is injected into the drill string–inner casing annulus through the inner-outer casing annulus and a port at the lower section of the inner-casing string. The inner-casing string can be retrieved when necessary. With this gas injection method, again only a portion of the cased hole section of the annulus is aerated. It allows for the use of conventional MWD, better pressure stability control at the bottomhole, and prevention of complications at the bottom of the inner-casing string.

10.2.2 Pressure Stability Control

As shown in Figure 10.7, making a pipe connection or changing a choke setting can cause severe fluctuation in the bottomhole pressure. Figure 10.8 illustrates the effect of the liquid injection rate on the pressure

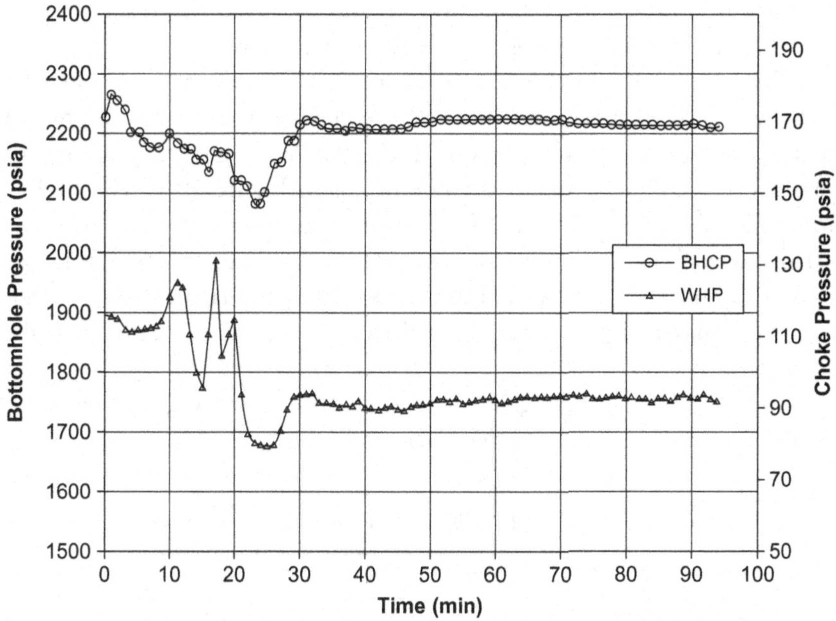

Figure 10.7 A field example showing pressure fluctuation during a pipe connection.

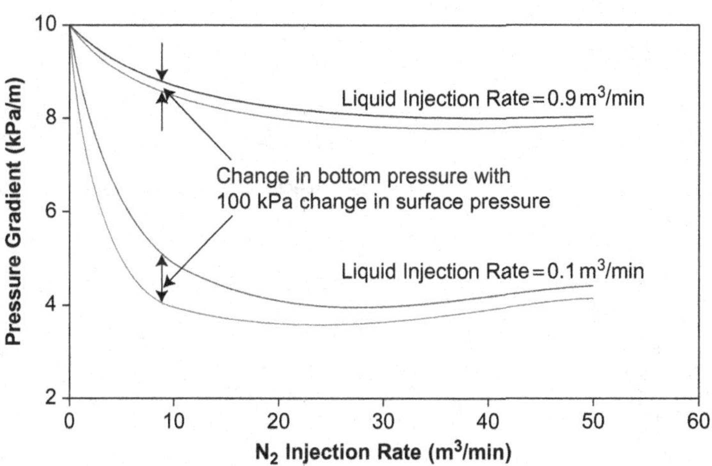

Figure 10.8 Effect of liquid injection rate on pressure fluctuation at the bottomhole.

fluctuation at the bottomhole. The pressure fluctuation in aerated liquid drilling has detrimental effects on both drilling operations and resultant well productivity. It can cause borehole collapse problems and excessive reservoir fluid influxes that need handling during drilling. It also induces formation damage due to the pressure surge inside the reservoir that mobilizes the formation fines and plugs reservoir pores.

The pressure fluctuation can be characterized with the term *instability factor*, which is defined as the ratio of the change in bottomhole pressure to the change in surface pressure ($\Delta P_{BH}/\Delta P_S$). Figure 10.9 plots the pressure instability factor for the conditions in Figure 10.8. The highest instability factor occurs for gas injection rates between 5 m^3/min and 10 m^3/min for the examined cases. Beyond this gas injection range, increasing the gas injection rate lowers the instability factor. It is generally believed that instability factors below 1.5 are manageable in UBD operations.

Guo and Ghalambor (2006) presented the following mathematical model for estimating the instability factor:

$$\frac{\Delta P_{BH}}{\Delta P_S} \approx \frac{dP_{BH}}{dP_S} = \frac{b + \dfrac{(1-2bm)(P_S+m) - (m + \frac{bn}{c} - bm^2)}{(P_S+m)^2 + n}}{b + \dfrac{(1-2bm)(P_{BH}+m) - (m + \frac{bn}{c} - bm^2)}{(P_{BH}+m)^2 + n}} \tag{10.1}$$

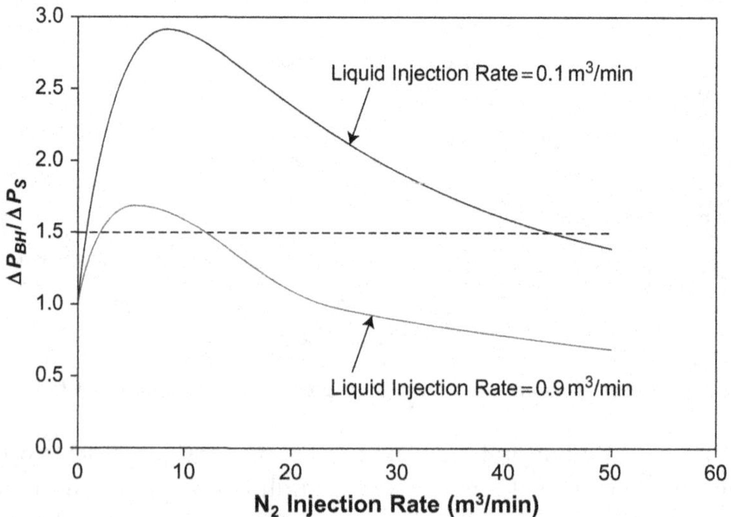

Figure 10.9 Pressure instability factor for the conditions in Figure 10.8.

where all the parameters are defined by Eqs. (9.4) through (9.10) in Chapter 9. Although the gas–liquid rate windows (GLRWs) can be prepared in the design stage, it is a good practice to optimize the combinations of gas and liquid flow rates in field operations. Such an optimization can be easily performed using simple computer programs such as *Pressure Instability.xls* in this book.

Illustrative Example 10.1

For the following given conditions, investigate the range of the pressure instability factor.

Total depth: 10,000 ft
Depth of the surface choke: 0 ft
Annulus OD: 6.28 in
Drill String OD: 3.5 in
Inclination angle: 0 deg
Surface temperature: 520 R
Rock specific gravity: 2.65 (water = 1)
Liquid weight: 8.4 ppg
Gas specific gravity: 0.97 (air = 1)
Formation fluid specific gravity: 0.8 (water = 1)
Geothermal gradient: 0.01°F/ft
Hole roughness: 0.0018 in
Formation fluid influx rate: 0 bbl/hr
Bit size: 6.125 in
Rate of penetration: 30 ft/hr
Liquid injection rate: 200 gpm
Gas injection rate: 500 scfm
Backpressure at choke: 50 psia

Solution

This problem can be solved with computer program *Pressure Instability.xls*. Figure 10.10 presents the calculated pressure instability factor along depth for three different liquid pumping rates while other parameters are fixed. It shows that the instability factor ranges from 1 to 2.5, and the higher the liquid flow rate, the lower the pressure instability factor. In order to achieve instability factor values below 1.5, the liquid injection rate should be above 200 gpm.

The effect of the gas injection rate on the instability factor is presented in Figure 10.11. It indicates that the higher the gas injection rate, the lower the pressure instability factor. However, for gas injection rates higher than 500 scfm, the instability factors at all depths are below 1.5.

(Continued)

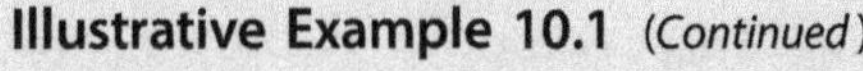

Illustrative Example 10.1 *(Continued)*

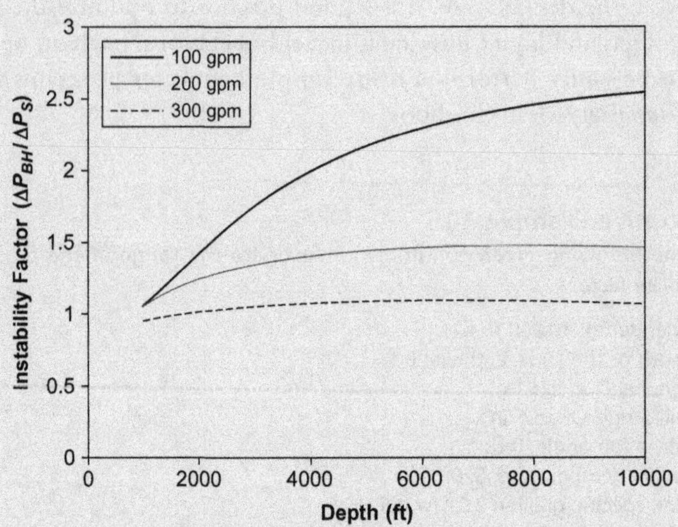

Figure 10.10 Effect of the liquid pumping rate on the instability factor. Gas injection rate 500 scfm, backpressure 50 psia.

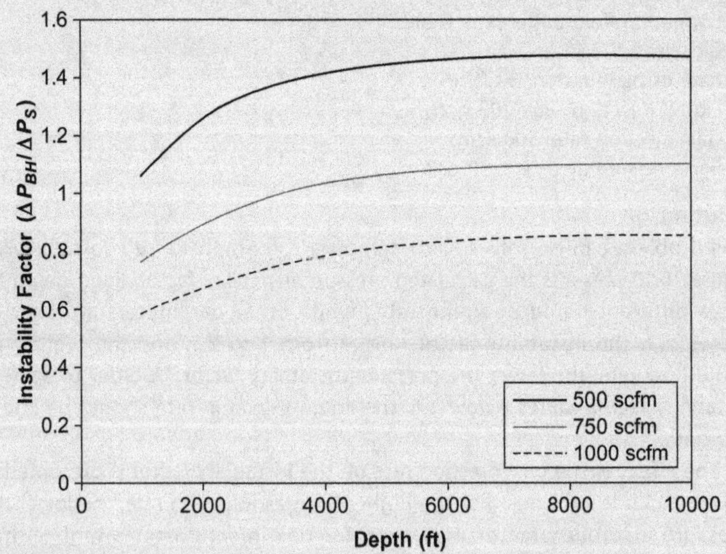

Figure 10.11 Effect of the gas injection rate on the instability factor. Gas injection rate 200 gpm, backpressure 50 psia.

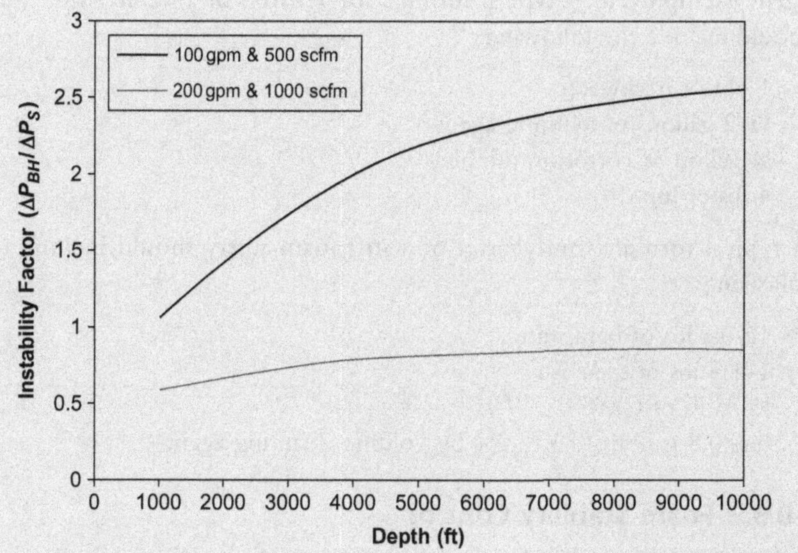

Figure 10.12 Effect of combination of liquid and gas flow rates on the instability factor.

Figure 10.12 shows the calculated pressure instability factor profiles for two different liquid–gas rate combinations while the injection gas–liquid ratio is kept the same. It indicates that it is the combination of the liquid and gas injection rates, not the injection GLR, that determines the pressure instability factor. Figures 10.10 through 10.12 show that the magnitude of the pressure instability factor increases with depth.

10.3 FOAM DRILLING

Foam has much better pressure stability. The major issues in foam drilling operations are foam stability in the borehole and foam handling at the surface. Foam stability is controlled by foam formulation, the gas–liquid ratio (GLR), and backpressure. Foam handling at the surface can be improved using closed-loop foam regeneration systems.

10.3.1 Foam Formulations

The best performance formulae of foams depend on the foaming agents (surfactants) used. Pilot tests are always required when a new foaming agent is employed. A typical formula for 1 barrel of a stable foam slurry should include the following:

- 1 bbl of freshwater
- ½–2 gallons of foaming agent
- ¼ gallon of corrosion inhibitor
- 4 lbs of lime

A typical formula for 1 barrel of a stiff foam slurry should include the following:

- 12–14 lbs of bentonite
- 1–1.5 lbs of soda ash
- ½–¾ lb of CMC
- 0.4–0.8 gallons (0.5 to 2% by volume) foaming agent

10.3.2 Foam Stability Control

It has been observed in laboratories that foams are stable when their gas contents are between 55% and 97.5% (Sanghani, 1982). The gas content is also called the foam quality index or simply the foam quality in foam drilling. It is defined as

$$\Gamma = \frac{\text{Gas Volume}}{\text{Total Foam Volume}} \tag{10.2}$$

or

$$\Gamma = \frac{\frac{4.07T}{P}Q_{go}}{\frac{4.07T}{P}Q_{go} + \frac{1}{7.48}Q_l + \frac{5.615}{60}Q_{fx}} \tag{10.3}$$

where T = temperature in °R; P = pressure in lb/ft^2; Q_{go} = gas injection rate in scfm; Q_l = liquid injection rate in gpm; and Q_{fx} = formation fluid influx rate in bbl/hour. Apparently, foam quality drops as the pressure increases with depth.

Foam drilling operations are usually designed with the maximum foam quality at the top hole being equal to 0.95 and the minimum foam quality at the bottomhole being equal to 0.60. However, these conditions are not maintained when the depth is beyond 5,000 ft. For deep drilling operations with foams, it is vitally important to use a high enough liquid injection rate for hole cleaning purposes because foam stability is not guaranteed in the lower section of the annulus. Guo and colleagues (2003) presented a method for foam stability control by adjusting the GLR and backpressure. It is summarized following.

Equation (10.3) can be rearranged as

$$\Gamma = \frac{\frac{4.07T}{P} GLR}{\frac{4.07T}{P} GLR + 0.09358\frac{Q_{fx}}{Q_l} + 0.13369} \quad (10.4)$$

If the backpressure is not applied, at the surface (flow line) condition, the foam quality reaches its highest value. For the foam quality not to exceed its maximum allowable value of Γ_{max}, the injection GLR should be controlled. Setting P and T to the atmospheric pressure and temperature, respectively, the maximum allowable injection GLR can be solved from Eq. (10.4) as

$$GLR_{max} = \frac{\Gamma_{max}}{(1-\Gamma_{max})}\left(0.09358\frac{Q_{fx}}{Q_l} + 0.13369\right) \quad (10.5)$$

Figure 10.13 presents a chart to determine the GLR_{max} for different values of Γ_{max} and Q_{fx}/Q_l. If this GLR_{max} results in a foam quality less than the minimum allowable value of foam quality Γ_{min} at the bottomhole, a gas–liquid ratio value of higher than the GLR_{max} should be used to achieve the minimum allowable value of foam quality Γ_{min} at the bottomhole. In this situation, backpressure should be applied to reduce the foam quality within its maximum allowable value of Γ_{max} at the surface. The minimum required backpressure can be solved from Eq. (10.4) as

$$P_{min} = \frac{4.07T(1-\Gamma_{max})GLR}{\Gamma_{max}\left(0.09358\frac{Q_{fx}}{Q_l} + 0.13369\right)} \quad (10.6)$$

Assuming $T = 520$ °R and $\Gamma_{max} = 0.975$, the minimum required backpressures are calculated and plotted in Figure 10.14 for different values of Q_{fx}/Q_l.

For a given foam drilling operation, the gas injection rate, and thus GLR, is limited. With this constraint of GLR, the foam quality at the bottomhole will decrease to its minimum allowable value of quality Γ_{min} when the bottomhole pressure increases to a critical value. This critical

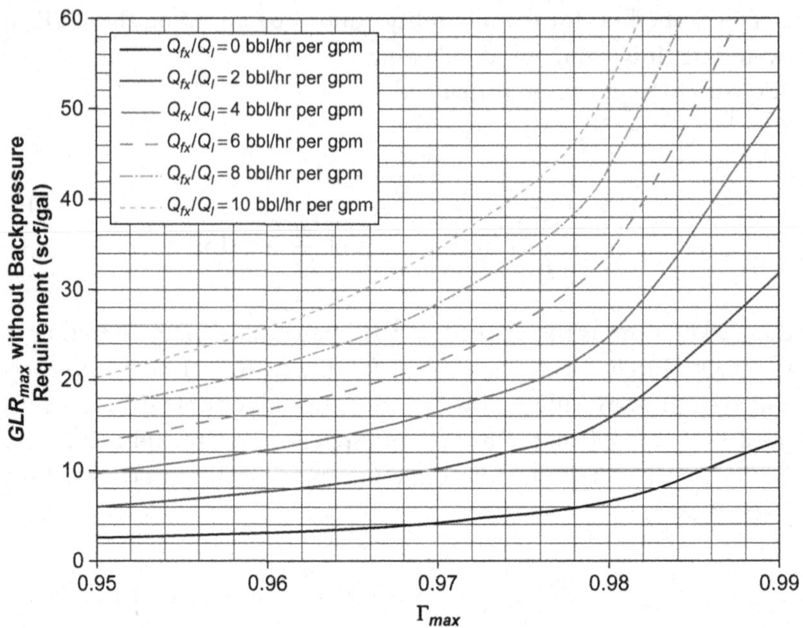

Figure 10.13 Effect of reservoir fluid influx on the maximum allowable injection GLR.

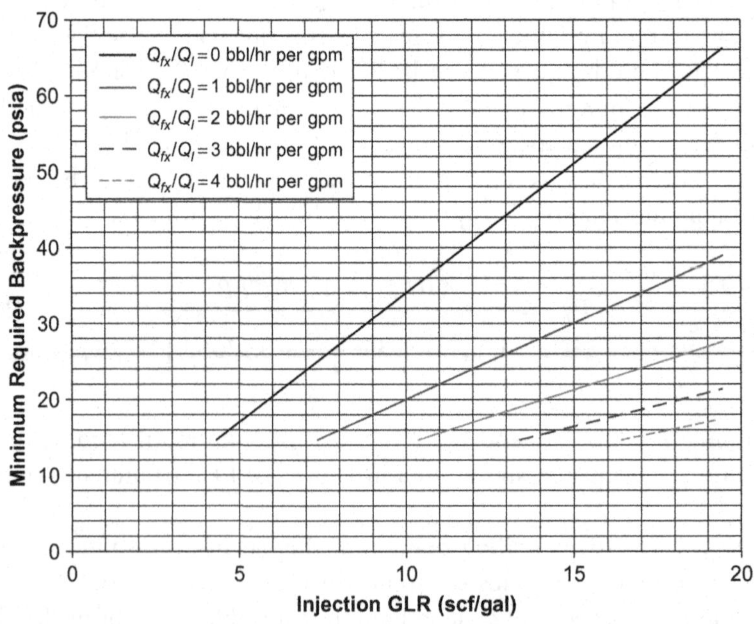

Figure 10.14 Effect of reservoir fluid influx on the minimum required backpressure.

bottomhole pressure is called the maximum allowable pressure for stable foam drilling, and it can be solved from Eq. (10.4):

$$P_{max} = \frac{4.07T(1-\Gamma_{min})GLR}{\Gamma_{min}\left(0.09358\frac{Q_{fx}}{Q_l}+0.13369\right)} \quad (10.7)$$

Assuming T = 520 °R and Γ_{min} = 0.55, the maximum pressures are calculated and plotted in Figure 10.15 for different values of Q_{fx}/Q_l. The figure indicates that the maximum pressure is less than 2,000 psia even with a GLR value of 20 scfm/gpm. This does not mean that wells cannot be drilled with foams when the bottomhole pressures are greater than 2,000 psia. In fact, many wells have been drilled with foams at bottomhole pressures higher than 2,000 psia. The explanation is that although the foams may not be stable at the bottomhole, hole cleaning can still be achieved with adequate mixture flow velocities in the annulus.

With the constraint of GLR, the foam quality at the bottomhole will decrease to its minimum allowable value of quality Γ_{min} at a critical depth.

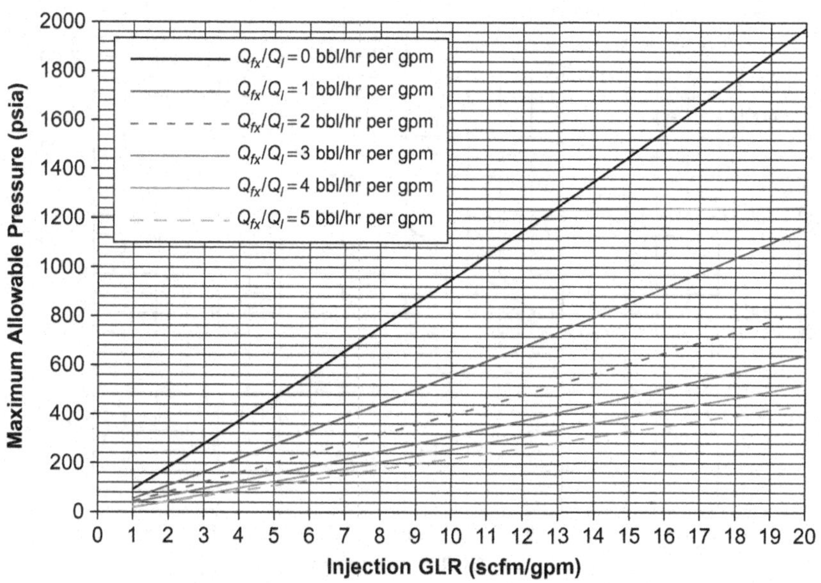

Figure 10.15 Effect of reservoir fluid influx on the maximum allowable pressure.

This critical depth is called the maximum depth for stable foam drilling, and it can be solved from Eq. (9.3) as

$$L_{max} = \frac{1}{a(1+d^2e)} \left\{ \begin{aligned} & b(P_{max} - P_{s-min}) + \frac{1-2bm}{2} \ln\left| \frac{(P_{max}+m)^2+n}{(P_{s-min}+m)^2+n} \right| \\ & - \frac{m+bn/c-bm^2}{\sqrt{n}} \left[\tan^{-1}\left(\frac{P_{max}+m}{\sqrt{n}}\right) - \tan^{-1}\left(\frac{P_{s-min}+m}{\sqrt{n}}\right) \right] \end{aligned} \right\} \tag{10.8}$$

Assuming $T = 520\ ^\circ R$, $\Gamma_{min} = 0.55$, and a 12.25-in × 6.325-in annulus, the maximum depths and corresponding equivalent circulating densities (ECDs) are calculated and plotted in Figure 10.16 for $Q_{fx} = 0$. The figure indicates that the maximum depth is less than 5,000 ft even with a GLR value of 20 scfm/gpm. Again, this does not mean that wells cannot be drilled with foams at depths greater than 5,000 ft. In fact, many wells have been drilled with foams at depths deeper than 5,000 ft. The explanation is

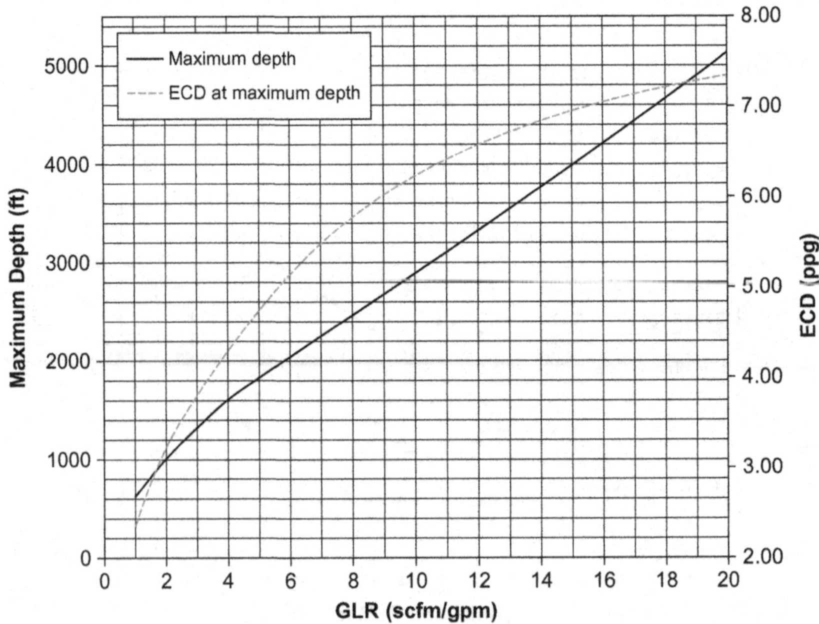

Figure 10.16 Effect of injection GLR on the maximum depth and ECD.

again that although the foams may not be stable at the bottomhole, hole cleaning can still be achieved with adequate mixture flow velocities in the annulus.

10.3.3 Handling the Returned Foam

Knowing how to handle the returned foam is important in foam drilling because it greatly affects drilling costs and impacts the environment. Returned foam can be either disposed or recycled. Foam disposal involves dumping the returned foam in an onsite pit if the environment permits. This option does not require special equipment. It may take a few days to a few weeks for stiff foams to decompose naturally. Therefore, in most cases, the returned foams are broken down with acids or separators before disposal. Engineers have known for years that acid added to a foam fluid will break down and destabilize the foam. In the late 1970s and early 1980s, techniques for breaking down foams using acid were developed to aid in cleaning up reservoir pits full of foam after foam drilling operations were completed. Lowering the pH of foam acts to destabilize it and allows the gas phase to break out much more easily.

Foam recycling requires special equipment. The process is illustrated in Figure 8.5. The procedure serves to greatly reduce the costs associated with using foam. It is a long process to continually foam, defoam, and then refoam the drilling fluid for reuse. In cases where containment or environmental concerns limit the use of foam because of the increased fluid volume, the foam recycling system eliminates these concerns by rapidly defoaming the fluid in the blooey line using a special defoamer. "Rapidly" means the foam half-life goes from 6 minutes to less than 15 seconds. The system results in a volume reduction of 95% within seconds. The system does not destroy the foaming agent.

The "defoamed" fluid can be cycled through the shale shaker to remove drill solids and then cycled through the steel pits for complete settling of the cuttings. An activator is then added to refoam the fluid. The activator completely counteracts the defoamer. It refoams to its original stability level. The number of foam, defoam, and refoam cycles is virtually unlimited. At pH levels above 10, the foaming agents, when agitated with water and air, create a viscous, stable foam. When the pH is dropped to 3.5 by the addition of sulfuric acid, the foaming potential of the water is greatly diminished. In fact, the foam is essentially killed by the addition of the acid.

SUMMARY

This chapter presented a brief introduction to the UBD operations with aerated liquids and foams. Pressure stability in aerated liquid drilling can be improved using annular injection of gas and increasing the mixture flow rate. Annular injection of gas also allows for conventional MWD to be used in directional drilling. Foam stability can be enhanced by adjusting injection GLR and backpressure. When the hole depth is greater than 5,000 ft, where foam is not expected to be stable, a higher mixture flow rate should be used for hole cleaning purposes.

REFERENCES

Guo, B., Ghalambor, A., 2006. Characterization and analysis of pressure instability in aerated liquid drilling. J. Can. Pet. Technol. (June).

Guo, B., Sun, K., Ghalambor, A., 2003. A closed form hydraulics equation for predicting bottom-hole pressure in UBD with foam. In: Paper SPE 81640, Proceedings of IADC/SPE Underbalanced Technology Conference and Exhibition, March 25–26, Houston.

Sanghani, V., 1982. Rheology of foam and its implications in drilling and cleanout operations. M.S. Thesis, University of Tulsa.

PROBLEMS

10.1 Solve Illustrative Example 10.1 assuming a 7⅞-in bit size and a 4½-in drill pipe OD.

10.2 Assuming $T = 560$ °R, reproduce Figure 10.13.

10.3 Assuming $T = 560$ °R, reproduce Figure 10.14.

10.4 Assuming $T = 560$ °R, reproduce Figure 10.15.

Unit Conversion Factors

Quantity	U.S. Field Unit	To SI Unit	To U.S. Field Unit	SI Unit
Length	feet (ft)	0.3084	3.2808	meter (m)
	mile (mi)	1.609	0.6214	kilometer (km)
	inch (in)	25.4	0.03937	millimeter (mm)
Mass	ounce (oz)	28.3495	0.03527	gram (g)
	pound (lb)	0.4536	2.205	kilogram (kg)
	lbm	0.0311	32.17	slug
Volume	gallon (gal)	0.003785	264.172	meter^3 (m^3)
	cu. ft (ft^3)	0.028317	35.3147	meter^3 (m^3)
	barrel (bbl)	0.15899	6.2898	meter^3 (m^3)
	Mcf (60°F, 4.7 psia)	28.317	0.0353	Mm^{3}15°C (101.325 kPa) (M = 1,000)
Area	acre	4.0469×10^3	2.471×10^{-4}	meter^2 (m^2)
	sq. ft (ft^2)	9.29×10^{-2}	10.764	meter^2 (m^2)
	sq. mile	2.59	0.386	(km)^2
Pressure	lb/in^2 (psi)	6.8948	0.145	kPa (1,000 Pa)
	psi	0.0680	14.696	atm
	psi/ft	22.62	0.0442	kPa/m
	inch (Hg)	3.3864×10^3	0.2953×10^{-3}	Pa
Temperature	Fahrenheit (°F)	0.5556 (°F – 32)	1.8C + 32	Celsius (°C)
	Rankine (°R)	0.5556	1.8	Kelvin (K)
Energy (work)	Btu	252.16	3.966×10^{-3}	cal
	Btu	1.0551	0.9478	kilojoule (kJ)
	ft-lbf	1.3558	0.73766	joule (J)
	hp-hr	0.7457	1.341	kW-hr
Viscosity	cp	0.001	1,000	Pa·s
	lb/ft-sec	1.4882	0.672	kg/(m-sec) or (Pa·s)
	lbf-s/ft^2	479	0.0021	dyne-s/cm^2 (poise)
Thermal conductivity	Btu-ft/hr-ft^2-°F	1.7307	0.578	W/(m·K)
Specific heat	Btu/(lbm-°F)	1	1	cal/(g-°C)
	Btu/(lbm-°F)	4.184×10^3	2.39×10^{-4}	J (kg-K)
Density	lbm/ft^3	16.02	0.0624	kg/m^3

Minimum Gas Flow Rates Required for Lifting Solids and Water in Air Drilling

Well geometry
Total measured depth: 400~1,5000 ft
Bit diameter: 4.75~9.875 in
Drill pipe OD: 2.375~5 in

Material properties
Specific gravity of rock: 2.75 (water = 1)
Specific gravity of gas: 1 (air = 1)
Gas specific heat ratio: 1.25
Specific gravity of oil: 1 (water = 1)
Specific gravity of formation water: 1.07 (water = 1)
Pipe roughness: 0.0018 in
Borehole roughness: 0.1 in

Environment
Site elevation: 0 ft
Ambient pressure: 14.7 psia
Ambient temperature: 60°F
Relative humidity: 0 fraction
Geothermal gradient: 0.01 F/ft

Operating conditions
Surface choke/flow line pressure: 14.7 psia
Rate of penetration: 60~120 ft/hour
Rotary speed: 50 rpm
Oil influx rate: 0 bbl/hour
Water influx rate: 4–24 bbl/hour
Interfacial tension: 40~60 dynes/cm

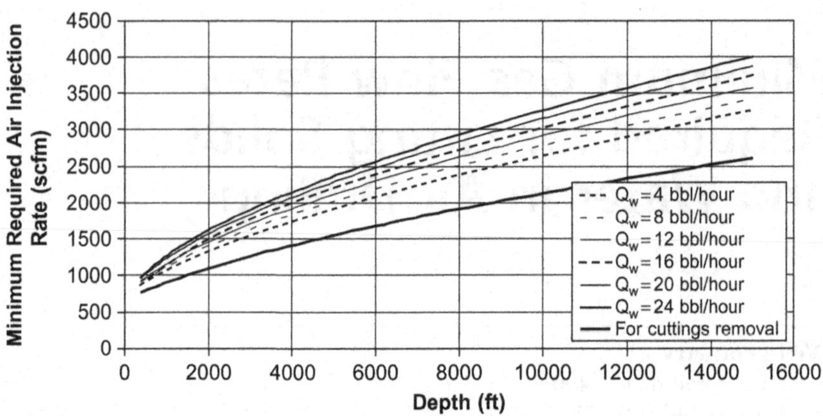

Figure B.1 Calculated air injection requirements for drilling a 7⅞-in hole with a 4½-in drill pipe at an ROP of 60 ft/hr (water-air interfacial tension = 60 dynes/cm).

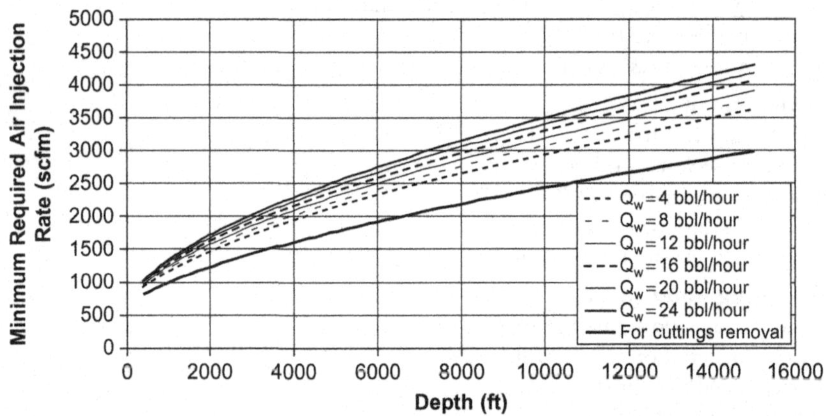

Figure B.2 Calculated air injection requirements for drilling a 7⅞-in hole with a 4½-in drill pipe at an ROP of 120 ft/hr (water-air interfacial tension = 60 dynes/cm).

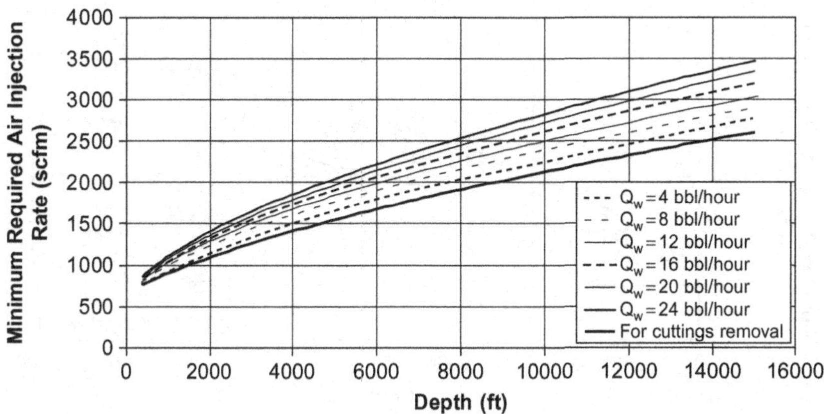

Figure B.3 Calculated air injection requirements for drilling a 7⅞-in hole with a 4½-in drill pipe at an ROP of 60 ft/hr (water-air interfacial tension = 40 dynes/cm).

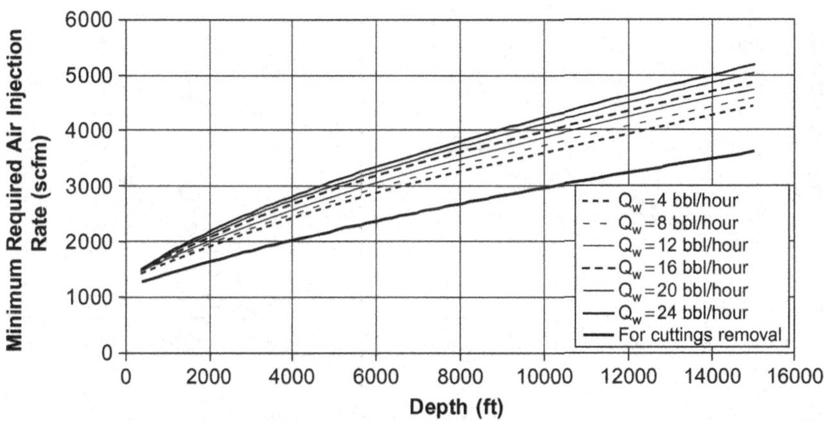

Figure B.4 Calculated air injection requirements for drilling a 9⅞-in hole with a 5-in drill pipe at an ROP of 60 ft/hr (water-air interfacial tension = 60 dynes/cm).

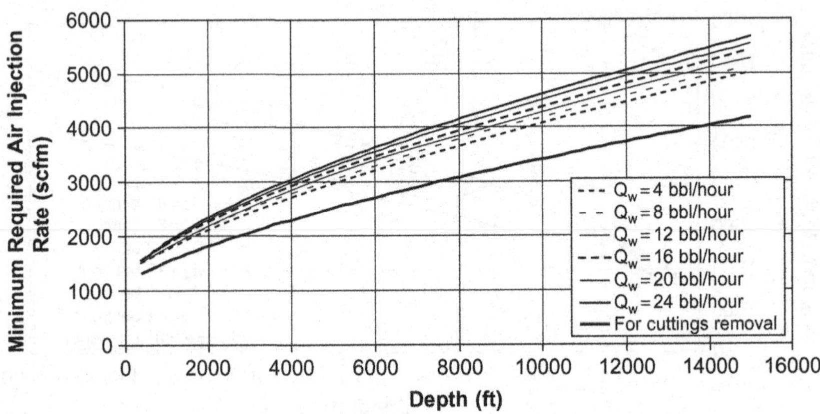

Figure B.5 Calculated air injection requirements for drilling a 9⅞-in hole with a 5-in drill pipe at an ROP of 120 ft/hr (water-air interfacial tension = 60 dynes/cm).

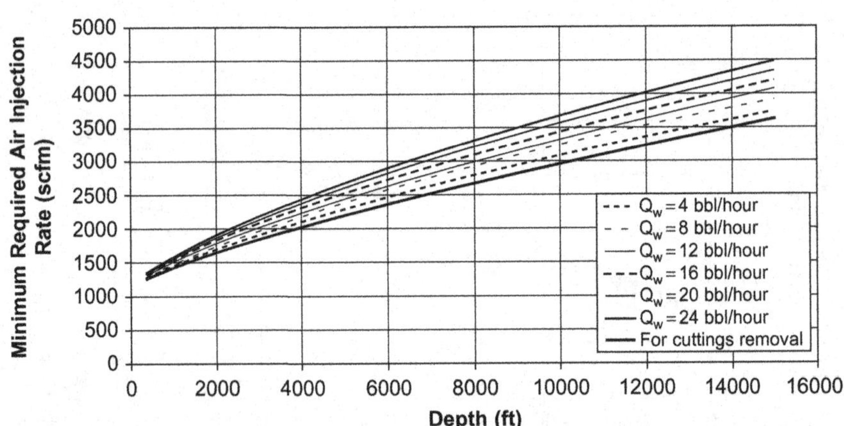

Figure B.6 Calculated air injection requirements for drilling a 9⅞-in hole with a 5-in drill pipe at an ROP of 60 ft/hr (water-air interfacial tension = 40 dynes/cm).

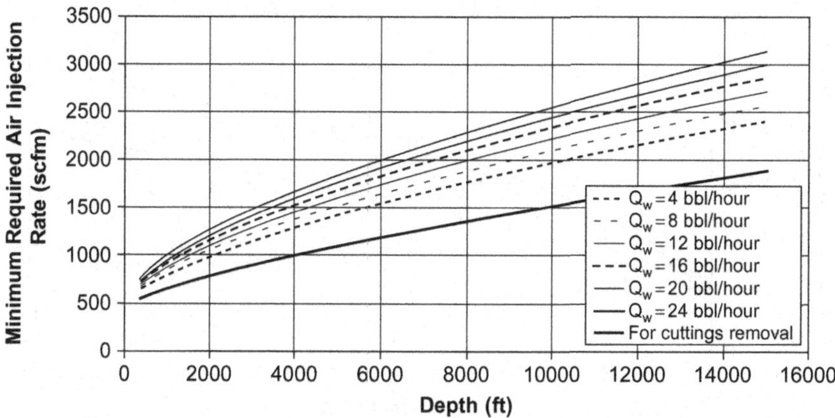

Figure B.7 Calculated air injection requirements for drilling a 6¼-in hole with a 2⅞-in drill pipe at an ROP of 60 ft/hr (water-air interfacial tension = 60 dynes/cm).

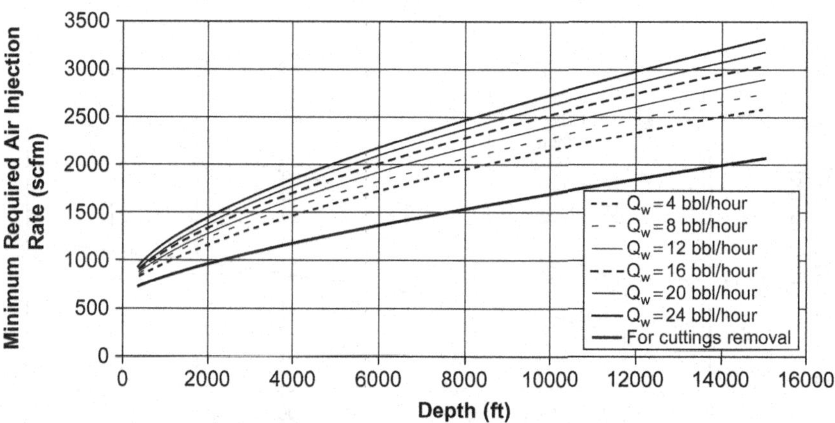

Figure B.8 Calculated air injection requirements for drilling a 6¼-in hole with a 2⅞-in drill pipe at an ROP of 120 ft/hr (water-air interfacial tension = 60 dynes/cm).

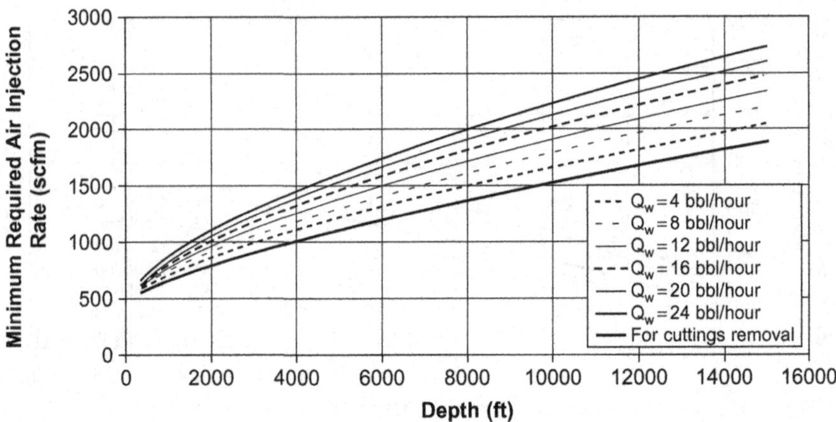

Figure B.9 Calculated air injection requirements for drilling a 6¼-in hole with a 2⅞-in drill pipe at an ROP of 60 ft/hr (water-air interfacial tension = 40 dynes/cm).

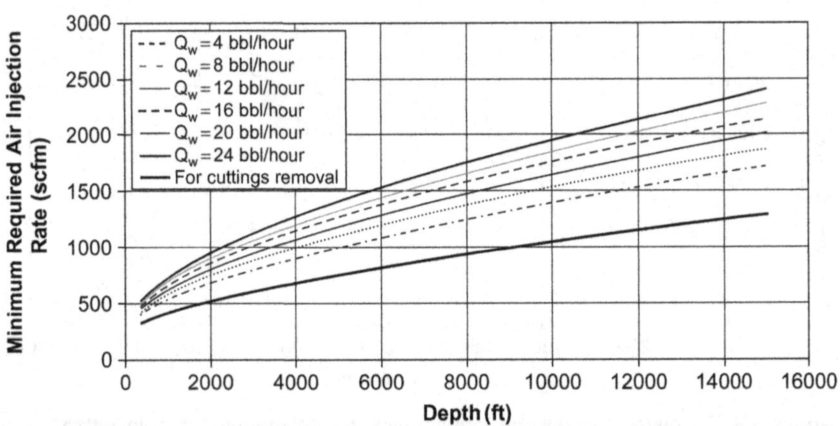

Figure B.10 Calculated air injection requirements for drilling a 4¾-in hole with a 2⅜-in drill pipe at an ROP of 60 ft/hr (water-air interfacial tension = 60 dynes/cm).

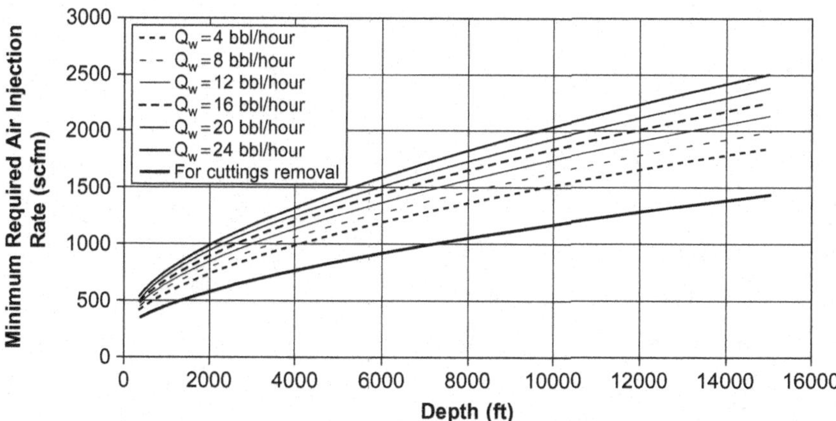

Figure B.11 Calculated air injection requirements for drilling a 4¾-in hole with a 2⅜-in drill pipe at an ROP of 120 ft/hr (water-air interfacial tension = 60 dynes/cm).

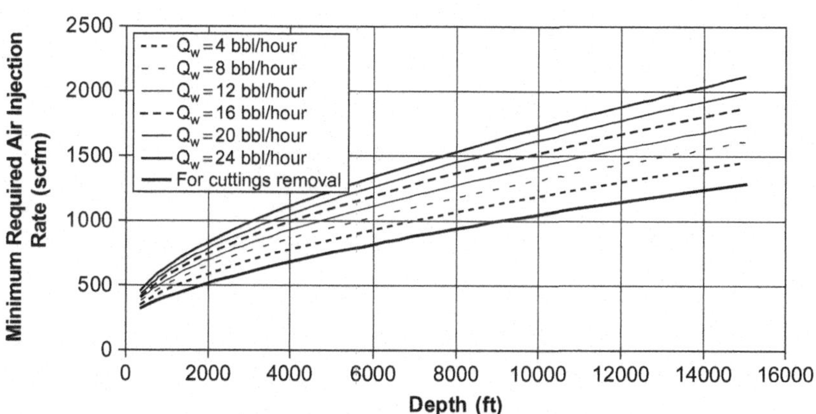

Figure B.12 Calculated air injection requirements for drilling a 4¾-in hole with a 2⅜-in drill pipe at an ROP of 60 ft/hr (water-air interfacial tension = 40 dynes/cm).

API Drill Collar Weight (lb/ft)

1	2	3	4	5	6	7	8	9	10	11	12	13	14
Drill Collar OD (in)	Drill Collar ID (in)												
	1	1¼	1½	1¾	2	2¼	2½	2¾	3	3¼	3½	3¾	4
2⅞	19	18	16										
3	21	20	18										
3⅛	22	22	20										
3¼	26	24	22										
3½	30	29	27										
3¾	35	33	32										
4	40	39	37	35	32	29							
4⅛	43	41	39	37	35	32							
4¼	46	44	42	40	38	35							
4½	51	50	48	46	43	41							
4¾			54	52	50	47	44						
5			61	59	56	53	50						
5¼			68	65	63	60	57						
5½			75	73	70	67	64	60					
5¾			82	80	78	75	72	67	64	60			
6			90	88	85	83	79	75	72	68			
6¼			98	96	94	91	88	83	80	76	72		
6½			107	105	102	99	96	91	89	85	80		
6¾			116	114	111	108	105	100	98	93	89		
7			125	123	120	117	114	110	107	103	98	93	84
7¼			134	132	130	127	124	119	116	112	108	103	93
7½			144	142	139	137	133	129	126	122	117	113	102
7¾			154	152	150	147	144	139	136	132	128	123	112
8			165	163	160	157	154	150	147	143	138	133	122
8¼			176	174	171	168	166	160	158	154	149	144	133
8½			187	185	182	179	176	172	169	165	160	155	150
9			210	208	206	203	200	195	192	188	184	179	174
9½			234	232	230	227	224	220	216	212	209	206	198
9¾			248	245	243	240	237	232	229	225	221	216	211
10			261	289	257	254	251	246	243	239	235	230	225
11			317	315	313	310	307	302	299	295	291	286	281
12			379	377	374	371	368	364	361	357	352	347	342

API Drill Pipe Dimensional Data

1	2	3
Pipe OD (in)	Nominal Weight (lb/ft)	Pipe ID (in)
2⅜	4.85	1.995
	6.65	1.815
2⅞	6.85	2.441
	10.4	2.151
3½	9.50	2.992
	13.30	2.764
	15.50	2.602
4	11.85	3.476
	14.00	3.340
	15.70	3.240
4½	13.75	3.958
	16.60	3.826
	20.00	3.640
	22.82	3.500
5	16.25	4.408
	19.50	4.276
	25.60	4.000
5½	19.20	4.892
	21.90	4.778
	24.70	4.670
6⅝	25.20	5.965
	27.20	5.901

API Casing Dimensional Data

1	2	3	4
Casing Outside Diameter (in)	Nominal Weight (lb/ft)	Grade	Inside Diameter (in)
4½	9.50	H-40	4.090
	9.50	J-55	4.090
	10.50	J-55	4.052
	11.60	J-55	4.000
	9.50	K-55	4.090
	10.50	K-55	4.052
	11.60	K-55	4.000
	11.60	C-75	4.000
	13.50	C-75	3.920
	11.60	L-80	4.000
	13.50	L-80	3.920
	11.60	N-80	4.000
	13.50	N-80	3.920
	11.60	C-90	4.000
	13.50	C-90	3.920
	11.60	C-95	4.000
	13.50	C-95	3.920
	11.60	P-110	4.000
	13.50	P-110	3.920
	15.10	P-110	3.826
	15.01	Q125	3.826
5	11.50	J-55	4.560
	13.00	J-55	4.494
	15.00	J-55	4.408
	11.50	K-55	4.560
	13.00	K-55	4.494
	15.00	K-55	4.408
	15.00	C-75	4.408
	18.00	C-75	4.276
	21.40	C-75	4.126
	23.20	C-75	4.044
	24.10	C-75	4.000
	15.00	L-80	4.408

(Continued)

1	2	3	4
Casing Outside Diameter (in)	Nominal Weight (lb/ft)	Grade	Inside Diameter (in)
5 *(cont'd)*	18.00	L-80	4.276
	21.40	L-80	4.126
	23.20	L-80	4.044
	24.10	L-80	4.000
	15.00	N-80	4.408
	18.00	N-80	4.276
	21.40	N-80	4.126
	23.20	N-80	4.044
	24.10	N-80	4.000
	15.00	C-90	4.408
	18.00	C-90	4.276
	21.40	C-90	4.044
	23.20	C-90	4.044
	24.10	C-90	4.000
	15.00	C-95	4.408
	18.00	C-95	4.276
	21.40	C-95	4.126
	23.20	C-95	4.044
	24.10	C-95	4.000
	15.00	P-110	4.408
	18.00	P-110	4.276
	21.40	P-110	4.126
	23.20	P-110	4.044
	24.10	P-110	4.000
	18.00	Q-125	4.276
	21.40	Q-125	4.126
	23.20	Q-125	4.044
	24.10	Q-125	4.000
5½	14.00	H-40	5.012
	14.00	J-55	5.012
	15.50	J-55	4.950
	17.00	J-55	4.892
	14.00	K-55	5.012
	15.50	K-55	4.950
	17.00	K-55	4.892
	17.00	C-75	4.892
	20.00	C-75	4.778
	23.00	C-75	4.670
	17.00	L-80	4.892

1	2	3	4
Casing Outside Diameter (in)	**Nominal Weight (lb/ft)**	**Grade**	**Inside Diameter (in)**
	20.00	L-80	4.778
	23.00	L-80	4.670
	17.00	N-80	4.892
	20.00	N-80	4.778
	23.00	N-80	4.670
	17.00	C-90	4.892
	20.00	C-90	4.778
	23.00	C-90	4.670
	26.00	C-90	4.548
	35.00	C-90	4.200
	17.00	C-95	4.892
	20.00	C-95	4.778
	23.00	C-95	4.670
	17.00	P-110	4.892
	20.00	P-110	4.778
	23.00	P-110	4.670
	23.00	Q-125	4.670
6⅝	20.00	H-40	6.049
	20.00	J-55	6.049
	24.00	J-55	5.921
	20.00	K-55	6.049
	24.00	K-55	5.921
	24.00	C-75	5.921
	28.00	C-75	5.791
	32.00	C-75	5.675
	24.00	L-80	5.921
	28.00	L-80	5.791
	32.00	L-80	5.675
	24.00	N-80	5.921
	28.00	N-80	5.791
	32.00	N-80	5.675
	24.00	C-90	5.921
	28.00	C-90	5.791
	32.00	C-90	5.675
	24.00	C-95	5.921
	28.00	C-95	5.791
	32.00	C-95	5.675
	24.00	P-110	5.921
	28.00	P-110	5.791

(*Continued*)

1	2	3	4
Casing Outside Diameter (in)	Nominal Weight (lb/ft)	Grade	Inside Diameter (in)
6⅝ *(cont'd)*	32.00	P-110	5.675
	32.00	Q-125	5.675
7	17.00	H-40	6.538
	20.00	H-40	6.456
	20.00	J-55	6.456
	23.00	J-55	6.366
	26.00	J-55	6.276
	20.00	K-55	6.456
	23.00	K-55	6.366
	26.00	K-55	6.276
	23.00	C-75	6.366
	26.00	C-75	6.276
	29.00	C-75	6.184
	32.00	C-75	6.094
	35.00	C-75	6.004
	38.00	C-75	5.920
	23.00	L-80	6.366
	26.00	L-80	6.276
	29.00	L-80	6.184
	32.00	L-80	6.094
	35.00	L-80	6.004
	38.00	L-80	5.920
	23.00	N-80	6.366
	26.00	N-80	6.276
	29.00	N-80	6.184
	32.00	N-80	6.094
	35.00	N-80	6.004
	38.00	N-80	5.920
	23.00	C-90	6.366
	26.00	C-90	6.276
	29.00	C-90	6.184
	32.00	C-90	6.094
	35.00	C-90	6.004
	38.00	C-90	5.920
	23.00	C-95	6.366
	26.00	C-95	6.276
	29.00	C-95	6.184
	32.00	C-95	6.094
	35.00	C-95	6.004

1	2	3	4
Casing Outside Diameter (in)	Nominal Weight (lb/ft)	Grade	Inside Diameter (in)
	38.00	C-95	5.920
	26.00	P-110	6.276
	29.00	P-110	6.184
	32.00	P-110	6.094
	35.00	P-110	6.004
	38.00	P-110	5.920
	35.00	Q-125	6.004
	38.00	Q-125	5.920
7⅝	24.00	H-40	7.025
	26.40	J-55	6.969
	26.40	K-55	6.969
	26.40	C-75	6.969
	29.70	C-75	6.875
	33.70	C-75	6.765
	39.00	C-75	6.625
	42.80	C-75	6.501
	45.30	C-75	6.435
	47.10	C-75	6.375
	26.40	L-80	6.969
	29.70	L-80	6.875
	33.70	L-80	6.765
	39.00	L-80	6.625
	42.80	L-80	6.501
	45.30	L-80	6.435
	47.10	L-80	6.375
	26.40	N-80	6.969
	29.70	N-80	6.875
	33.70	N-80	6.765
	39.00	N-80	6.625
	42.80	N-80	6.501
	45.30	N-80	6.435
	47.10	N-80	6.375
	26.40	C-90	6.969
	29.70	C-90	6.875
	33.70	C-90	6.765
	39.00	C-90	6.625
	42.80	C-90	6.501
	45.30	C-90	6.435
	47.10	C-90	6.375

(*Continued*)

1	2	3	4
Casing Outside Diameter (in)	**Nominal Weight (lb/ft)**	**Grade**	**Inside Diameter (in)**
7⅝ *(cont'd)*	26.40	C-95	6.969
	29.70	C-95	6.875
	33.70	C-95	6.765
	39.00	C-95	6.625
	42.80	C-95	6.501
	45.30	C-95	6.435
	47.10	C-95	6.375
	29.70	P-110	6.875
	33.70	P-110	6.765
	39.00	P-110	6.625
	42.80	P-110	6.501
	45.30	P-110	6.435
	47.10	P-110	6.375
	39.00	Q-125	6.625
	42.80	Q-125	6.501
	45.30	Q-125	6.435
	47.10	Q-125	6.375
8⅝	28.00	H-40	8.017
	32.00	H-40	7.921
	24.00	J-55	8.097
	32.00	J-55	7.921
	36.00	J-55	7.825
	24.00	K-55	8.097
	32.00	K-55	7.921
	36.00	K-55	7.825
	36.00	C-75	7.825
	40.00	C-75	7.725
	44.00	C-75	7.625
	49.00	C-75	7.511
	36.00	L-80	7.825
	40.00	L-80	7.725
	44.00	L-80	7.625
	49.00	L-80	7.511
	36.00	N-80	7.825
	40.00	N-80	7.725
	44.00	N-80	7.625
	49.00	N-80	7.511
	36.00	C-90	7.825
	40.00	C-90	7.725

1	2	3	4
Casing Outside Diameter (in)	**Nominal Weight (lb/ft)**	**Grade**	**Inside Diameter (in)**
	44.00	C-90	7.625
	49.00	C-90	7.511
	36.00	C-95	7.825
	40.00	C-95	7.725
	44.00	C-95	7.625
	49.00	C-95	7.511
	40.00	P-110	7.725
	44.00	P-110	7.625
	49.00	P-110	7.511
	49.00	Q125	7.511
9⅝	32.30	H-40	9.001
	36.00	H-40	8.921
	36.00	J-55	8.921
	40.00	J-55	8.835
	36.00	K-55	8.921
	40.00	K-55	8.835
	40.00	C-75	8.835
	43.50	C-75	8.755
	47.00	C-75	8.681
	53.50	C-75	8.535
	40.00	L-80	8.835
	43.50	L-80	8.755
	47.00	L-80	8.681
	53.50	L-80	8.535
	40.00	N-80	8.835
	43.50	N-80	8.755
	47.00	N-80	8.681
	53.50	N-80	8.535
	40.00	C-90	8.835
	43.50	C-90	8.755
	47.00	C-90	8.681
	53.50	C-90	8.535
	40.00	C-95	8.835
	43.50	C-95	8.755
	47.00	C-95	8.681
	53.50	C-95	8.535
	43.50	P-110	8.755
	47.00	P-110	8.681
	53.50	P-110	8.535

(Continued)

1	2	3	4
Casing Outside Diameter (in)	**Nominal Weight (lb/ft)**	**Grade**	**Inside Diameter (in)**
9⅝ *(cont'd)*	47.00	Q125	8.681
	53.50	Q125	8.535
10¾	32.75	H-40	10.192
	40.50	H-40	10.050
	40.50	J-55	10.050
	45.50	J-55	9.950
	51.00	J-55	9.850
	40.50	K-55	10.050
	45.50	K-55	9.950
	51.00	K-55	9.850
	51.00	C-75	9.850
	55.50	C-75	9.760
	51.00	L-80	9.850
	55.50	L-80	9.760
	51.00	N-80	9.850
	55.50	N-80	9.760
	51.00	C-90	9.850
	55.50	C-90	9.760
	51.00	C-95	9.850
	55.50	C-95	9.760
	51.00	P-110	9.850
	55.50	P-110	9.760
	60.70	P-110	9.660
	65.70	P-110	9.560
	60.70	Q-125	9.660
	65.70	Q-125	9.560
11⅝	42.00	H-40	11.084
	42.00	J-55	11.000
	54.00	J-55	10.880
	60.00	J-55	10.772
	47.00	K-55	11.000
	54.00	K-55	10.880
	60.00	K-55	10.772
	60.00	C-75	10.772
	60.00	L-80	10.772
	60.00	N-80	10.772
	60.00	C-90	10.772
	60.00	C-95	10.772

1	2	3	4
Casing Outside Diameter (in)	**Nominal Weight (lb/ft)**	**Grade**	**Inside Diameter (in)**
	60.00	P-110	10.772
	60.00	Q-125	10.772
13⅜	48.00	H-40	12.715
	54.50	J-55	12.615
	61.00	J-55	12.515
	68.00	J-55	12.415
	54.50	K-55	12.615
	61.00	K-55	12.515
	68.00	K-55	12.415
	68.00	C-75	12.415
	72.00	C-75	12.347
	68.00	L-80	12.415
	72.00	L-80	12.347
	68.00	N-80	12.415
	72.00	N-80	12.347
	68.00	G-90	12.415
	72.00	G-90	12.347
	68.00	C-95	12.415
	72.00	C-95	12.347
	68.00	P-110	12.415
	72.00	P-110	12.347
	72.00	Q-125	12.347
16	65.00	H-40	15.250
	75.00	J-55	15.124
	84.00	J-55	15.010
	75.00	K-55	15.124
	84.00	K-55	15.010
18⅝	87.50	H-40	17.755
	87.50	J-55	17.755
	87.50	K-55	17.755
20	94.00	H-40	19.124
	94.00	J-55	19.124
	106.50	J-55	19.000
	133.00	J-55	18.730
	94.00	K-55	19.124
	106.50	K-55	19.000
	133.00	K-55	18.730

INDEX

Note: Page numbers in *italics* indicate figures and tables.

Printed in the United States
By Bookmasters